CW01645216

INDEPENDENT LEARNING PROJECT FOR ADVANCED CHEMISTRY

ILPAC

ADVANCED PRACTICAL CHEMISTRY RESOURCE PACK

second edition

REVISED BY ANN LAINCHBURY JOHN STEPHENS ALEC THOMPSON

JOHN MURRAY

PREFACE

This *Resource Pack* for teachers is designed to accompany *Advanced Practical Chemistry*, which is a collection of all the ILPAC experiments (Independent Learning Project in Advanced Chemistry). The experiments cover the requirements of all the main GCE A-level syllabuses and should also be suitable for other courses, such as University Foundation Science.

The *Resource Pack* contains notes and advice on the experiments, specimen results, answers to questions, Technician's Sheets and blank tables, which can be photocopied (preferably enlarged) and issued to students to help them organise and simplify their work.

Some of the experiments have been identified as suitable for practical assessment, and for these we have provided suggested mark schemes.

The Technician's Sheets list all the requirements for each experiment in detail, together with comprehensive hazard warnings.

The use of the material in this pack should greatly simplify the organisation of practical work for GCE A-level in schools and colleges.

© John Murray (Publishers) Ltd 1997
Original material produced by the Independent Learning Project for Advanced Chemistry, sponsored by the Inner London Education Authority and published as ILPAC volumes S1–S4, P1–P5, O1–O4, I1–I6 and ILPAC Teacher's and Technician's Notes 1–6 and 7–12.
Taken from the second edition, published as ILPAC volumes 1–12 with additional material by Ann Lainchbury, John Stephens and Alec Thompson.

First edition published in 1985
by John Murray (Publishers) Ltd
50 Albemarle Street
London W1X 4BD

Reprinted 1987, 1992, 1994.
Second edition 1997.

All rights reserved. The material in this publication is copyright but permission is given for teachers to make copies of the photocopiable masters on pages 238–254 for one-time use as instructional material within their own school (or other educational institution). This permission does not extend to the making of copies for use outside the institution in which they are made (e.g. in a resource centre), and the material may not be copied in unlimited quantities, kept on behalf of others, passed on or sold to third parties or stored for future use in a retrieval system. If you wish to use the material in any way other than as specified you must apply in writing to the publishers.

Design by John Townson/Creation.
Layouts by Wearset, Boldon, Tyne and Wear.
Illustrations by Barking Dog Art.

Typeset in 10/12pt Times and Helvetica.
Printed and bound in Great Britain by St Edmundsbury Press, Bury St Edmunds.

A catalogue record for this book is available from the British Library.

ISBN 0 7195 7508 7

CONTENTS

International hazard symbols

Corrosive

Oxidising

Explosive

Radioactive

Harmful or irritant

Toxic

Highly flammable

INTRODUCTION TO TEACHER'S NOTES

COSHH regulations and risk assessments

The Control of Substances Hazardous to Health (COSHH) Regulations 1988 and the Management of Health and Safety Regulations 1992 require that a risk assessment is carried out before any hazardous activity, including the use of any substance known to be hazardous to health, is undertaken in any place of employment. While it is the employer's responsibility to ensure that risk assessments are carried out, the task in a chemistry department will be performed by teachers and technicians. We strongly recommend that, if you are not already clear about your responsibilities in this area, you discuss the matter with your head of science and/or principal who will clarify things with your employer if necessary. We cannot cover the topic fully here but useful discussions appear in *Education in Science* (Nov. 1994) and *School Science Review* (Dec. 1992), both published by The Association for Science Education. Other references are given in the Introduction to the Technician's Sheets (page 127).

A risk assessment requires the teacher and technician to consider the hazards to be met in a particular experiment and make decisions about the precautions to be taken. They must of course then be passed on to the pupils.

All of the experiments in *Advanced Practical Chemistry* have already been risk assessed. In each experiment the hazards are identified and the precautions to be taken are stated both in the students' book and in the technician's sheets.

CLEAPSS/ASE Laboratory Safeguards Committee has read the text and confirms that, in the draft it was asked to check, the identification of hazards and the precautions given conform with the published general assessments which most employers accept for risk assessments. The revision editors are grateful for the assistance and advice offered by the Committee. However, there is a continual need for staff to be alert to any changes in risk assessment circulated by their employers.

Also, although we give specific advice about certain substances and procedures, some basic precautions are assumed, as detailed below in the section 'Good practice with chemicals', written by CLEAPSS. These precautions would normally be covered in pre-A-level work, but you may be teaching students from a variety of backgrounds and, in any case, we recommend that you remind your students of these basic precautions at the start of their A-level study and encourage them to treat **all** chemicals and experimental procedures with care **as a matter of course**. A general discussion on safety, perhaps with some demonstrations, might be appropriate.

Good practice with chemicals

A brief statement such as this can only be a summary.

Eye protection should be worn whenever there is a recognised risk to the eyes. That includes activities involving heating, those involving corrosive, toxic or flammable chemicals, or those where heat may be generated in a chemical reaction.

Many accidents occur during heating activities. Long hair should be tied back, and ties, cardigans, etc. should not be allowed to hang freely. It is assumed that teachers will show and remind students how to heat safely solids in test-tubes and liquids in boiling-tubes, using small quantities so that the tube is not more than about one-fifth full, and pointing the tube away from their own and other people's faces.

It is assumed that teachers will show students how to smell the contents of a test-tube or bottle safely, by wafting the vapours towards their nose from some distance, without inhaling deeply and without pointing the test-tubes at their face.

It is assumed that students will be shown how to shake a test-tube safely in order to mix the liquid contents, recognising that this is most readily achieved if the tube is no more than one-quarter full.

It is assumed that students will be shown how to pour safely from bottles, pouring away from the label. Spillages of chemicals should always be wiped up; some spillages may require chemical treatment (e.g. neutralisation) but most can be dealt with by a damp cloth. Students should be trained to use a spatula, or similar device, and never to handle chemicals with their fingers. Work in schools rarely requires protective gloves (although they are occasionally recommended in this text). However, when chemicals have been used, students should be trained to wash their hands afterwards.

If the risk assessment requires the use of a fume cupboard, then this should be one which is at least to current standards (e.g. Design Note 29). If safety screens are required, for example in a demonstration, then these should be of sufficient number to protect the teacher and all the students.

It is expected that every school will have rules governing the behaviour of students in the laboratory. No eating (or indeed smoking or the application of cosmetics) should be allowed in laboratories.

The Introduction to the Technician's Sheets (page 127) gives details of HAZCARDS and a list of recommended books on laboratory management and safety.

Practical assessment

The internal assessment of course work is based on an assessment by the teacher, of practical skills relating to laboratory experiments carried out by the students. The abilities to be assessed vary slightly between the different boards but can generally be divided into the following four categories.

A. Planning and design
B. Manipulation
C. Observing and recording
D. Interpretation

Categories B and C are combined by some boards as a 'carrying out' skill.

Abilities are assessed on either a five-point or ten-point scale, depending on the particular syllabus.

We have included mark schemes for a number of the ILPAC experiments which can be used or adapted for use in the internal assessment of your practical work. The means used by different schools for ensuring comparability of assessment across the whole school will of course vary, however, and it is important to choose a system that will work well within your school. This will give a valid ranking order and indicate quantitative differences between students in each assessment set. Once submitted, the board will moderate either on a sample of the students' work (which includes work from top, middle and bottom of the distribution range) and accompanying mark schemes, or on the basis of performance on parts of the written examination paper. Your particular syllabus will give you details of the type of moderation your board will use.

The experiments for which we give suggested mark schemes are as follows:

Experiment 1.6	A titration exercise	skills B and D
Experiment 2.2	Determining an enthalpy change of solution	skill A
Experiment 3.5	Variation of boiling point with composition of a mixture	skill B
Experiment 4.9	Balancing a redox reaction	skill B
Experiment 4.10	Observation and deduction exercise 1	skills C and D
Experiment 6.3	Determining a solubility product	skills B and D
Experiment 7.6	Obtaining the pH curve for an acid–alkali titration	skills C and D
Experiment 7.7	Preparation of buffers	skill A
Experiment 7.13	Rusting of iron and steel	skills C and D
Experiment 8.2	Chemical properties of phenol	skills C and D

Experiment 8.7	Observation and deduction exercise 3	skills C and D
Experiment 9.6	Determining the activation energy of a reaction	skills B and D
Experiment 9.7	Determining the activation energy of a catalysed reaction	skill A
Experiment 10.3	Preparation of aspirin	skill B
Experiment 10.4	Identification of two organic compounds	skills C and D
Experiment 10.8	Identification of three organic compounds	skills C and D
Experiment 10.9	Preparing phenyl benzoate	skill B
Experiment 11.1	The reactions of tin and lead and their aqueous ions	skills B and C
Experiment 11.2	The preparations and reactions of tin(IV) oxide and lead(IV) oxide	skill B
Experiment 11.3	Observation and deduction exercise 4	skills C and D
Experiment 11.11	Observation and deduction exercise 5	skills C and D
Experiment 12.5	Observation and deduction exercise 6	skills C and D

Teachers might wish to adapt other experiments to make them suitable for assessment.

We use a five-point scale for assessment but this can easily be adapted to a ten-point scale if required.

Notes on experiments

Hazards in organic experiments

Before the students start the experiments in *Advanced Practical Chemistry* (particularly Experiments 5.1 to 5.4, and 8.1 to 8.7) we recommend a group discussion on the hazards of using organic compounds. Since the list of dangerous or suspect chemicals is growing all the time, we think that students should be advised to treat **all** organic compounds as potentially dangerous, even though this may mean treating some chemicals with more care than is necessary. We suggest the following guidelines.

Many organic compounds are:

- **flammable**, so **always stopper the bottles after use and remove them to a safe place away from flames**,
- **toxic** if inhaled, so **always handle organic chemicals in a fume cupboard** (with the extractor switched on, of course),
- **toxic** if absorbed through the skin even though there may be no immediate effect, so **always wear gloves**,
- **violent** in some of their reactions, so **always wear safety spectacles and pull down the front of the fume cupboard** (leaving a gap just large enough for the extractor to work efficiently).

The CLEAPSS development group at Brunel University have printed a set of HAZCARDS which not only state the potential dangers of many chemicals but also tell you what to do in the event of an accident. The most recent edition includes methods of disposal. They also produce sheets on the use of halogenated compounds and their alternatives (see below). It is a good idea for the technician to issue a HAZCARD with each chemical. HAZCARDS and other useful material are available from:

CLEAPSS School Science Service
Brunel University
Uxbridge UB8 3PH
Tel. 01895 251496

A qualified science teacher should always be in the laboratory throughout these experiments.

Disposal of organic residues

Contaminated residues may be disposed of as follows.

- **Up to 10 cm^3 at a time** Small quantities may be poured down a sink in a fume cupboard with plenty of flowing water and the addition of some detergent. If detergent is not added to some of the oily organic residues they may stick to the sides of the sink where they can then give off harmful vapour into the room. Residues from different experiments, containing different chemicals should not be mixed.
- **Larger volumes** You will need to consult your local education authority on the facility available for disposing of larger volumes, especially halogenoalkanes. Highly flammable solvents should **not** be put down the sewer in large amounts.
- **Organic solids** Small amounts can be put in the waste. If they are toxic then dilution of the solid with sand will render the mixture unpalatable (the authorities are concerned that children may put them in their mouths).

It is always advisable to consult your local education authority or water authority. Further details can be found in laboratory manuals such as CLEAPSS Laboratory Handbook. A full list of recommended books is given on page 129.

Use of halogenated compounds and their alternatives

Since January 1995, it has not been possible to obtain 1,1,1-trichloroethane or tetrachloromethane as both are known to be ozone-damaging chemicals. CLEAPSS has suggested some possible substitutes where applicable throughout this book. In this book we recommend you use a new solvent called **Volasil 244** in place of 1,1,1-trichloroethane as the solvent for bromine. It has other applications in other experiments. Volasils are non-toxic compounds which have no smell and are made for extensive use in the cosmetics industry where they act as solvents, for example in hair-care products they improve wet combing yet evaporate without a residue. Chemically they are cyclic silicones with four or five siloxane units. Volasil 244 solidifies in the bottle on a cold day. The bottle should be placed in warm water to melt the solid. Solutions of bromine in Volasil 244 are stable over a few weeks but not months. Volasil 244 is available from:

BDH Laboratory Supplies, Merck Ltd, Hunter Boulevard, Lutterworth LE17 4XN
Tel. 0800 223344
Costs (1997) are £12.80 for 500 ml and £27.10 for 2.5 litres.

Molecular models

A set of Molymod molecular models containing both rigid and flexible plastic bonds is available from: Spiring Enterprises Ltd, Beke Hall, Marringdean Road, Billingshurst, West Sussex RH14 9HF

TEACHER'S NOTES

EXPERIMENT 1.1 Determining the Avogadro constant

Hair seems to give better results than cotton and does not need initial greasing. It is important to make sure that the loop is in contact with the surface throughout its circumference – otherwise the oleic acid will leak out.

The students should obtain a value for L to within one power of 10 of the accepted value. You might ask more able students to investigate quantitatively the effect on the calculation of making various assumptions about the shape of the molecule.

Specimen results and calculations

Results Table 1.1

Number of drops to deliver 1 cm^3 of solution	Number of drops delivered to make monomolecular layer	Diameter of monomolecular layer/cm
100	12	11.2

1. $$\text{Volume of 1 drop} = \frac{1.0\ cm^3}{100} = 0.010\ cm^3$$

2. Since 1000 cm^3 of solution contains 0.050 cm^3 of oleic acid

$$1.00\ cm^3 \text{ of solution contains } \frac{0.050\ cm^3}{1000} \text{ of oleic acid}$$

$$\therefore 0.010\ cm^3 \text{ of solution contains } \frac{0.050\ cm^3 \times 0.010}{1000} \text{ of oleic acid} = 5.0 \times 10^{-7}\ cm^3$$

3. Volume of oleic acid in monomolecular layer = $5.0 \times 10^{-7}\ cm^3 \times 12 = 6.0 \times 10^{-6}\ cm^3$

4. $$A = \frac{3.142 \times (11.2\ cm)^2}{4} = 98.5\ cm^2$$

5. $$\text{Thickness} = \frac{\text{volume}}{\text{area}} = \frac{6.0 \times 10^{-6} cm^3}{98.5\ cm^2} = 6.1 \times 10^{-8}\ cm$$

6. Volume of one molecule = $(6.1 \times 10^{-8}\ cm)^3 = 2.3 \times 10^{-22}\ cm^3$

7. $$\text{Density} = \frac{\text{molar mass}}{\text{molar volume}}$$

$$\therefore \text{molar volume} = \frac{\text{molar mass}}{\text{density}} = \frac{282\ g\ mol^{-1}}{0.890\ g\ cm^{-3}} = 317\ cm^3\ mol^{-1}$$

8. $$L = \frac{\text{molar volume}}{\text{volume of molecule}} = \frac{317\ cm^3\ mol^{-1}}{2.3 \times 10^{-22}\ cm^3} = \mathbf{1.4 \times 10^{24}\ mol^{-1}}$$

(The accepted value of $L = 6.02 \times 10^{23}\ mol^{-1}$. Students should expect to obtain a value of L to within one power of 10, i.e. between $6.0 \times 10^{22}\ mol^{-1}$ and $6.0 \times 10^{24}\ mol^{-1}$.)

Answers to questions

1. The various sources of error which may account for the difference between the experimental and actual value for *L* are as follows:
   a the number of drops required to make a monomolecular layer was inaccurately measured;
   b not all the pentane had evaporated from the surface;
   c the volume of the drop was inaccurately determined;
   d the molecules were unevenly dispersed on the surface of the film, i.e. there was more than one layer of molecules on parts of the film;
   e the assumption about the shape of the molecule was incorrect. In fact, the molecule is rather more the shape of a cylinder than a cube.
2. The number of drops required to fill the loop is subject to the greatest error. Even if students are sure to the nearest drop, this gives only two significant figures (or only

one for a small loop!) whereas all the other values are obtained to three significant figures.

3. A substitute for pentane must:
   a dissolve oleic acid readily;
   b evaporate readily;
   c not react with oleic acid;
   d not react with water;
   e not dissolve in water.

EXPERIMENT 1.2 Preparing a standard solution

The students will need somewhere to keep their standard solutions until they do Experiment 1.3 – perhaps a safe corner in the prep. room. They must check that their solutions are clearly labelled and that the labels are firmly attached to the flasks.

Specimen results

Results Table 1.2

Molar mass of potassium hydrogenphthalate, M	204.1 g mol^{-1}
Mass of bottle and contents before transfer, m_1	15.47 g
Mass of bottle and contents after transfer, m_2	10.20 g
Mass of potassium hydrogenphthalate, $m = (m_1 - m_2)$	5.27 g
Amount of potassium hydrogenphthalate, $n = m/M$	2.58×10^{-2} mol
Volume of solution, V	0.250 dm^3
Concentration of potassium hydrogenphthalate, $c = n/V$	0.103 mol dm^{-3}

Answer to question

a The concentration would be lower than calculated. From the expression $c = n/V$, a decrease in amount will reduce the value of the concentration.

b The concentration would be greater than calculated. From the expression $c = n/V$, insufficient water (i.e. a decrease in V) will increase the value of the concentration.

EXPERIMENT 1.3 An acid–base titration

In order to check the students' accuracy, either you or your technician should standardise the sodium hydroxide solution used in the experiment. Since this is the students' first titration, you should watch them fairly closely and discuss any errors in their technique.

Specimen results

Results Table 1.3

Pipette solution	potassium hydrogenphthalate		0.0103	mol dm^{-3}	25.0 cm^3	
Burette solution	sodium hydroxide		?	mol dm^{-3}		
Indicator	phenolphthalein					
		Trial	1	2	3	(4)
Burette readings	Final	26.9	26.80	27.45	26.15	—
	Initial	0.7	0.90	1.50	0.30	—
Volume used (titre)/cm^3		26.2	25.90	25.95	25.85	—
Mean titre/cm^3		25.9(0)				

Calculation Let A refer to potassium hydrogenphthalate and B to sodium hydroxide. Substituting into the expression

$$\frac{c_A V_A}{c_B V_B} = \frac{a}{b}$$

where $c_A = 0.103\ \text{mol dm}^{-3}$ $\quad c_B = ?$
$V_A = 25.0\ \text{cm}^3$ $\quad V_B = 25.9\ \text{cm}^3$
$a = 1$ $\quad b = 1$

gives

$$\frac{0.103\ \text{mol dm}^{-3} \times 25.0\ \text{cm}^3}{c_B \times 25.9\ \text{cm}^3} = \frac{1}{1}$$

$$\therefore\ c_B = \frac{0.103\ \text{mol dm}^{-3} \times 25.0\ \text{cm}^3}{25.9\ \text{cm}^3} = \mathbf{0.0994\ mol\ dm^{-3}}$$

Answers to questions

1. **a** If the burette is dry (and clean), no effect. If the burette is wet, this will slightly dilute the NaOH so that slightly more of it is required. If V_B is increased, c_B will be **smaller** than it should be.
   b If the pipette is dry (and clean), no effect. If the pipette is wet, this will slightly dilute the acid so that slightly less NaOH is required. If V_B is decreased, c_B will be **greater** than it should be.
   c The recorded volume of NaOH will be greater than it should be, because it will include a bubble of air. If V_B is increased, c_B will be **smaller** than it should be.
   d No effect. The dilution occurs **after** the amount has been measured.
2. It is easier for the eye to detect the approach of the end-point when colour appears rather than when it disappears.
3. Sodium hydroxide solution attacks glass to some extent, especially the ground surfaces in some burette taps. Also, it absorbs carbon dioxide from the air, forming a crust of sodium carbonate. Both these actions can cause burette taps to seize up.

EXPERIMENT 1.4 A redox titration

This is a straightforward experiment which should give good results. Freshly prepared starch solution is required for a sharp end-point.

Specimen results
Results Table 1.4

Pipette solution	iodine			0.0497 mol dm^{-3}		10.0 cm^3
Burette solution	sodium thiosulphate			0.0512 mol dm^{-3}		
Indicator	starch					
		Trial	1	2	3	(4)
Burette readings	Final	20.3	40.40	20.15	40.25	—
	Initial	0.0	20.30	0.10	20.15	—
Volume used (titre)/cm^3		20.3	20.10	20.05	20.10	—
Mean titre/cm^3		20.1(0)				

Calculation

1. Substituting into the expression

$$\frac{c_A V_A}{c_B V_B} = \frac{a}{b}$$ ('A' refers to $Na_2S_2O_3$, 'B' to I_2)

where $c_A = 0.0512\ \text{mol dm}^{-3}$ $\quad c_B = 0.0497\ \text{mol dm}^{-3}$

$V_A = 20.1\ \text{cm}^3$ $\quad V_B = 10.0\ \text{cm}^3$

$a = ?$ $\quad b = ?$

gives $$\frac{a}{b} = \frac{0.0512\ \text{mol dm}^{-3} \times 20.1\ \text{cm}^3}{0.0497\ \text{mol dm}^{-3} \times 10.0\ \text{cm}^3} = \frac{2.07}{1}$$

$\therefore$ $a = 2$ and $b = 1$. So we can write for the equation:

$$2Na_2S_2O_3(aq) + I_2(aq) \rightarrow \text{Products}$$

2. The formula for the other compound is $Na_2S_4O_6$ (sodium tetrathionate).

$$2Na_2S_2O_3(aq) + I_2(aq) \rightarrow NaI(aq) + ?$$

To balance I atoms, the stoichiometric coefficient of NaI must be 2.

$$2Na_2S_2O_3(aq) + I_2(aq) \rightarrow 2NaI + ?$$

The atoms unaccounted for are two of sodium, four of sulphur and six of oxygen, $\mathbf{Na_2S_4O_6}$. The balanced equation is

$$2Na_2S_2O_3(aq) + I_2(aq) \rightarrow 2NaI(aq) + Na_2S_4O_6(aq)$$

(The empirical formula is NaS_2O_3, but $Na_2S_4O_6$ is preferred because it contains $S_4O_6^{2-}$ ions.)

EXPERIMENT 1.5 A precipitation titration

The students are given a standard solution of silver nitrate. Since AR silver nitrate has a purity of at least 99.9%, a standard solution can be prepared by direct weighing. If, however, commercial recrystallised silver nitrate is used, the prepared solution should be standardised with pure sodium chloride using potassium chromate indicator.

Some students may experience difficulty in detecting the end-point of this titration, i.e. the point at which the pale yellow colour **just** changes to a faint reddish-brown colour. These students should be advised to prepare a blank at the start of the experiment by mixing together about 10.0 cm^3 of the barium chloride solution, 1 g of sodium sulphate, 5.0 cm^3 of the silver nitrate solution and a few drops of indicator. The end-point of their titrations can then be taken to be the point at which the colour of their titration mixture departs from the pale colour of the blank.

Warn your students not to waste or discard silver nitrate solution in view of its prohibitive cost. See that they return the contents of their titration-flasks to the silver residues bottle, so that the silver may be recovered and silver nitrate regenerated.

Specimen results

Results Table 1.5

Mass of bottle and contents before transfer, m_1	11.79 g
Mass of bottle and contents after transfer, m_2	10.21 g
Mass of sample, $m = (m_1 - m_2)$	1.58 g
Mass of $BaCl_2{\cdot}xH_2O$ in 10.0 cm^3 of solution	0.0632 g

Results Table 1.6

Pipette solution	barium chloride			? mol dm^{-3}		10.0 cm^3
Burette solution	silver nitrate			0.0506 mol dm^{-3}		
Indicator	starch					
		Trial	1	2	3	(4)
Burette readings	Final	10.4	20.65	30.90	41.10	—
	Initial	0.0	10.40	20.75	30.90	—
Volume used (titre)/cm^3		10.4	10.25	10.15	10.20	—
Mean titre/cm^3		10.2(0)				

Calculation

There is more than one way of tackling this problem. Here is one suggestion.

First calculate the masses of $BaCl_2$ and H_2O in each sample. The mass of $BaCl_2$ can be calculated indirectly from titration data and the mass of H_2O by difference.

Next convert masses to relative amounts to determine *x*. We explain this step by step.

1. Amount of Cl in sample (as Cl^-)

$$Ag^+(aq) + Cl^-(aq) \rightarrow AgCl(s)$$

$$\therefore \text{ amount of Cl} = \text{amount of } AgNO_3$$
$$= cV = 0.506 \text{ mol dm}^{-3} \times 0.0102 \text{ dm}^3$$
$$= 5.16 \times 10^{-4} \text{ mol}$$

2. Amount and mass of $BaCl_2$

$$BaCl_2(s) + aq \rightarrow Ba^{2+}(aq) + 2Cl^-(aq)$$

$$\therefore \text{ amount of } BaCl_2 = \tfrac{1}{2} \times \text{amount of } Cl^-$$
$$= \tfrac{1}{2} \times 5.16 \times 10^{-4} \text{ mol} = 2.58 \times 10^{-4} \text{ mol}$$
$$\text{mass of } BaCl_2 = nM = 2.58 \times 10^{-4} \text{ mol} \times 208 \text{ g mol}^{-1}$$
$$= 0.0537 \text{ g}$$

3. Mass and amount of H_2O

$$\text{mass of } H_2O = \text{mass of } BaCl_2{\cdot}xH_2O - \text{mass of } BaCl_2$$

$$= 0.0632 \text{ g} - 0.0537 \text{ g} = 0.0095 \text{ g}$$

$$\text{amount of } H_2O = \frac{m}{M} = \frac{0.0095 \text{ g}}{18.0 \text{ g mol}^{-1}} = 5.3 \times 10^{-4} \text{ mol}$$

4. Relative amounts

	$BaCl_2$	H_2O
Amount/mol	2.58×10^{-4}	5.3×10^{-4}
Amount/ smallest amount = relative amount	$\frac{2.58 \times 10^{-4}}{2.58 \times 10^{-4}} = 1.00$	$\frac{5.3 \times 10^{-4}}{2.58 \times 10^{-4}} = 2.1$

The relative amounts are very close to the integers 1 and 2.

$\therefore$ $x = 2$ and the formula is $\mathbf{BaCl_2{\cdot}2H_2O}$.

EXPERIMENT 1.6 A titration exercise

Here is an opportunity to test your students' techniques and accuracy in titration. If your particular syllabus requires practical work to be internally assessed, then you could use this as one of your assessed practicals.

In this practical, the students titrate ammonium iron(II) sulphate solution against standard potassium manganate(VII) in order to determine the number of molecules of water of hydration in ammonium iron(II) sulphate crystals, i.e. determine x in the formula $FeSO_4(NH_4)_2SO_4{\cdot}xH_2O$.

Neither potassium manganate(VII) nor ammonium iron(II) sulphate are primary standards. However, it is sufficient for this experiment to prepare the solutions by direct weighing provided that AR grade chemicals are used and care is taken to avoid oxidation of the iron(II) by air.

Suggested mark scheme

You should perform this titration yourself in order to set a standard against which the students' titration values may be compared. When assessing accuracy, we suggest you give full marks for a titration value which agrees with the standard (teacher's mean titration value) to $\pm 0.10\ cm^3$, with decreasing marks down to $\pm 0.30\ cm^3$. No marks for accuracy should be given for errors greater than $\pm 0.40\ cm^3$. Where consistent but inaccurate titres are obtained, use your judgement. Some errors cannot be blamed on the student.

Total marks for manipulation = 10, as follows:

Correct (safe) filling of burette	(1)
Correct use of pipette filler	(1)
Correct use of pipette	(1)
Good titration technique with dropwise addition near the end-point	(1)
Carries out an approximate and at least two accurate titrations	(1)
Accuracy of mean titration	(3)
Agreement of accurate titrations to $\pm 0.1\ cm^3$	(2)

Divide total mark by 2 to obtain scaled mark.

Total marks for interpretation = 5, as follows:

Calculation of $[Fe^{2+}]$ from titration result	(2)
Calculation of mass of anhydrous $Fe(NH_4)_2(SO_4)_2$ in 1 dm^3 of solution	(1)
Calculation of mass of water of crystallisation	(1)
Calculation of value of x	(1)

Scaled mark = total mark.

EXPERIMENT 1.7 Using a hand spectroscope to observe emission spectra

Students should work in pairs because it is not easy to see many of the spectral lines in flame colours due to their low intensity and short duration. You may prefer, if it is possible, to borrow a set of discharge tubes and relevant apparatus from a physics department – the lines are much easier to see. In any case, you should make sure the students have access to some coloured plates, or transparencies, of emission spectra to help them to recognise the lines they see. A good plate can be found as a centre-fold in the *Revised Nuffield Science Book of Data*. A useful alternative to the flame test routine is to hold with tongs a filter paper soaked in a solution of the chloride.

Answer to question

A continuous spectrum shows a continuous range of wavelengths. In the visible range, this appears as broad bands of colour merging into one another with no sharp boundaries.

An emission spectrum shows a limited number of wavelengths, which appear as distinct lines. In the visible range, the lines appear in different colours.

EXPERIMENT 2.1 Determining an enthalpy change of reaction

It is worth trying out this experiment to see that your sample of zinc reacts at a suitable rate. If the zinc is very finely powdered, or if it is in too great excess, the reaction occurs so rapidly that the point of plotting the temperature/time curve is lost. Conversely, coarse zinc in the stoichiometric amount may not react completely in a reasonable time so that the maximum temperature cannot be determined accurately.

Students should get excellent results – low results may be due to impure zinc (partially oxidised). Students may notice a side reaction, hydrogen being evolved from the acidic $CuSO_4$(aq), but this is small in extent and has a similar value for ΔH°_r, so that little error is introduced.

Specimen results and calculation

Results Table 2.1

Time/min	0.0	0.5	1.0	1.5	2.0	2.5	3.0	3.5	4.0	4.5
Temperature/°C	27.0	27.0	27.2	27.2	27.2	27.2	–	66.0	71.4	71.8
Time/min	5.0	5.5	6.0	6.5	7.0	7.5	8.0	8.5	9.0	9.5
Temperature/°C	70.2	68.0	66.2	64.4	62.8	61.0	59.5	58.0	56.6	55.1

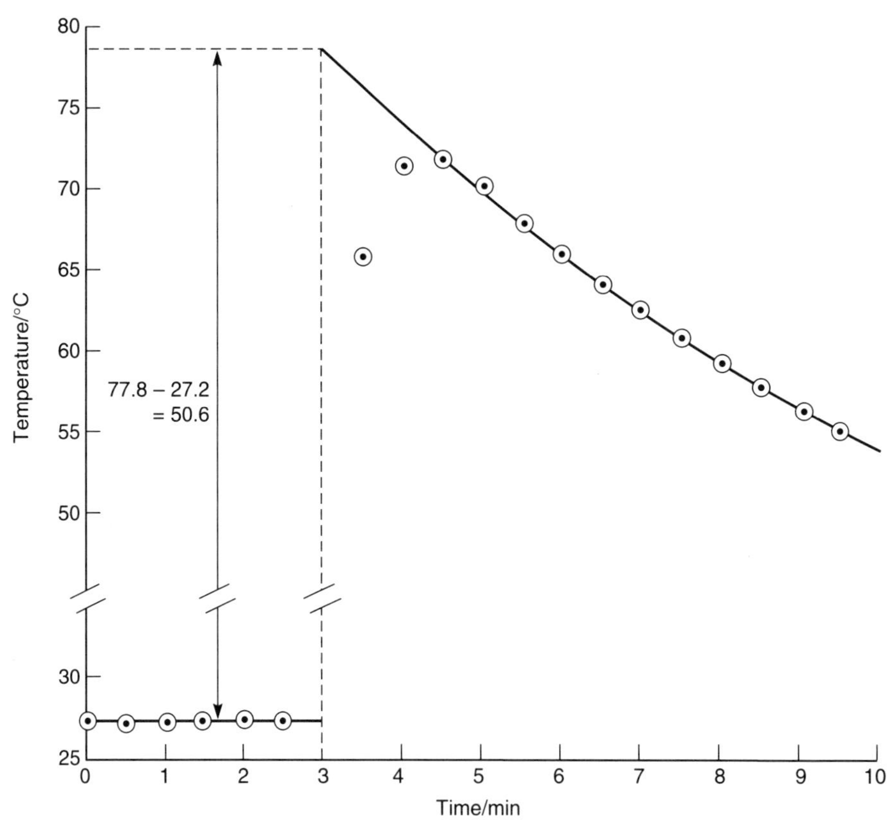

[enthalpy change due to reaction (at constant T)] = [enthalpy change in restoring solution to original temperature]

i.e. $\Delta H = mc_p\Delta T$

$\therefore \Delta H = mc_p\Delta T = 0.0250 \text{ kg} \times 4.18 \text{ kJ kg}^{-1} \text{ K}^{-1} \times (-50.6 \text{ K}) = -5.29 \text{ kJ}$

Amount of Cu^{2+} used $= cV = 1.00 \text{ mol dm}^{-3} \times 0.0250 \text{ dm}^3 = 0.0250 \text{ mol}$

Scaling up to the amount in the equation, 1 mol,

$$\Delta H = -5.29 \text{ kJ} \times \frac{1}{0.0250} = \mathbf{-212 \text{ kJ}}$$

$$\text{Zn(s)} + \text{Cu}^{2+}\text{(aq)} \rightarrow \text{Cu(s)} + \text{Zn}^{2+}\text{(aq)};\ \Delta H^{\circ} = -212 \text{ kJ mol}^{-1}$$

Answers to questions

1.
$$\text{Error} = \frac{-212-(-217) \text{ kJ mol}^{-1}}{-217 \text{ kJ mol}^{-1}} \times 100 = \frac{+5}{-217} \times 100 = \mathbf{-2.3\%}$$

2. **a** There is some heat loss from the polystyrene cup.
   b The heat capacity of the solution is not precisely 4.18 kJ kg^{-1} K^{-1}.
   c The density of the solution is not precisely 1.00 g dm^{-3}
   d The heat capacities of the metals were ignored.
   e The thermometer has a small but measurable heat capacity.
3. The reaction takes a few minutes to complete because the copper first precipitated shields some of each zinc particle from the Cu^{2+}.

EXPERIMENT 2.2 Determining an enthalpy change of solution

This provides a good opportunity for a practical assessment on 'planning skills', as required by some examining boards. The experiment is very simple; requirements and procedure are similar to those in Experiment 2.1. However, a temperature/time curve is not necessary because the temperature change is much smaller and occurs very rapidly.

Since the temperature change is small, students should use a thermometer graduated in 0.1 °C. Suitable masses are 2.67 g of NH_4Cl (0.05 mol) and 90.0 g of H_2O – avoid too small a scale of operation. Students should observe a temperature drop of nearly 2 K, giving a value of ΔH°_{soln} within 2 or 3 kJ mol^{-1} of the accepted value of 16.4 kJ mol^{-1}.

Suggested mark scheme

Total marks for planning = 10, as follows:
Requirements list (2)
Calculation of masses required, as below (other quantities are acceptable provided they are in the correct molar ratio of 1:100 and are suitable for the apparatus) (2)
Indication of procedure: weigh (or measure volume of) water, put water in polystyrene cup, measure steady temperature of water, weigh the salt, add salt to water and stir to dissolve, record greatest temperature change. One mark each (6)

Total marks	10–9	8–7	6–5	4–3	2–1	1–0
Scaled mark	5	4	3	2	1	0

Specimen results and calculations

Suitable amounts are 0.0500 mol of NH_4Cl and 5.00 mol of H_2O.
Then mass of NH_4Cl = 53.4 g mol^{-1} × 0.0500 mol = 2.67 g
and mass of H_2O = 18.0 g mol^{-1} × 5.00 mol = 90.0 g.

$$\Delta T = \text{final temperature} - \text{initial temperature} = 22.7\ °\text{C} - 24.5\ °\text{C} = -1.8 \text{ K}$$
$$\therefore \Delta T \text{ for transfer between system and surroundings} = +1.8 \text{ K}$$
$$\Delta H = mc_p\Delta T = 0.090 \text{ kg} \times 4.18 \text{ kJ kg}^{-1} \text{ K}^{-1} \times 1.8 \text{ K} = +0.68 \text{ kJ}$$

Scaling up from 0.050 mol to 1 mol:

$$\Delta H = +\ 0.068 \text{ kJ} \times (1/0.050) = \mathbf{+14 \text{ kJ}}$$

$$\text{NH}_4\text{Cl(s)} + 100\text{H}_2\text{O(l)} \rightarrow \text{NH}_4\text{Cl(aq, 100H}_2\text{O)};\ \Delta H^{\circ} = +14 \text{ kJ mol}^{-1}$$

Note that only two significant figures are justified because ΔT is small.

Answer to question

Results are usually lower than +16.4 kJ mol^{-1}, largely because some heat energy enters the system to increase the final temperature a little. It is difficult to measure a small temperature change accurately and an error of only 0.1 °C changes the final result by almost 1 kJ mol^{-1}. Also, we have made the usual assumptions about heat capacities of the polystyrene cup (= 0), thermometer (= 0) and solution (= that of water).

EXPERIMENT 2.3 Using Hess' law (1)

The choice of magnesium sulphate over the more commonly used copper sulphate enables good results to be obtained in simple apparatus because the temperature changes are greater and the salt more readily dehydrated. However, if you wish your students to use a more sophisticated apparatus and method, you can give them Experiment 2.6 as an alternative. In both experiments, some care should be taken to obtain, and keep dry, samples of the anhydrous salts. A sensitive thermometer is essential.

Specimen results and calculations

Results Table 2.2

	$MgSO_4$	$MgSO_4{\cdot}7H_2O$
Mass of weighing bottle	12.91 g	13.42 g
Mass of weighing bottle + salt	15.92 g	19.58 g
Mass of salt	3.01 g	6.16 g
Mass of polystyrene cup	2.10 g	2.36 g
Mass of polystyrene cup + water	47.10 g	44.21 g
Mass of water	45.00 g	41.85 g
Initial temperature	24.1 °C	24.8 °C
Final temperature	35.4 °C	23.4 °C

1. $\Delta H = -mc_p\Delta T = -(0.0450\text{ kg} \times 4.18\text{ kJ kg}^{-1}\text{ K}^{-1} \times 11.3\text{ K}) = -2.13\text{ kJ}$

   Scaling up, $\Delta H = -2.13\text{ kJ} \times \dfrac{1\text{ mol}}{0.0250\text{ mol}} = -85.0\text{ kJ}$

   Thus $MgSO_4(s) + 100H_2O(l) \rightarrow MgSO_4(aq, 100H_2O)$; $\Delta H^\ominus = \mathbf{-85.0\ kJ\ mol^{-1}}$

2. $\Delta H = -mc_p\Delta T = -(0.04185\text{ kg} \times 4.18\text{ kJ kg}^{-1}\text{ K}^{-1} \times (-1.4\text{ K}) = +0.24\text{ kJ}$

   Scaling up, $\Delta H = +0.24\text{ kJ} \times \dfrac{1\text{ mol}}{0.0250\text{ mol}} = +9.8\text{ kJ}$

   Thus $MgSO_4{\cdot}7H_2O(s) + 93H_2O(l) \rightarrow MgSO_4(aq, 100H_2O)$; $\Delta H^\ominus = \mathbf{+9.8\ kJ\ mol^{-1}}$

3.

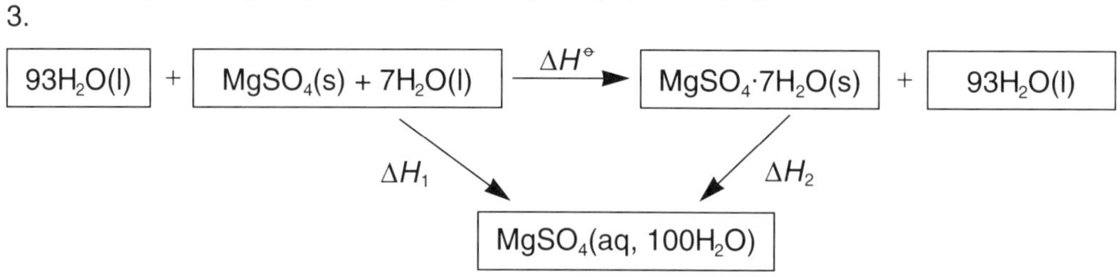

$\Delta H^\ominus = \Delta H_1 - \Delta H_2 = (-85.0 - 9.8)\text{ kJ mol}^{-1} = \mathbf{-94.8\ kJ\ mol^{-1}}$

Answers to questions

1.

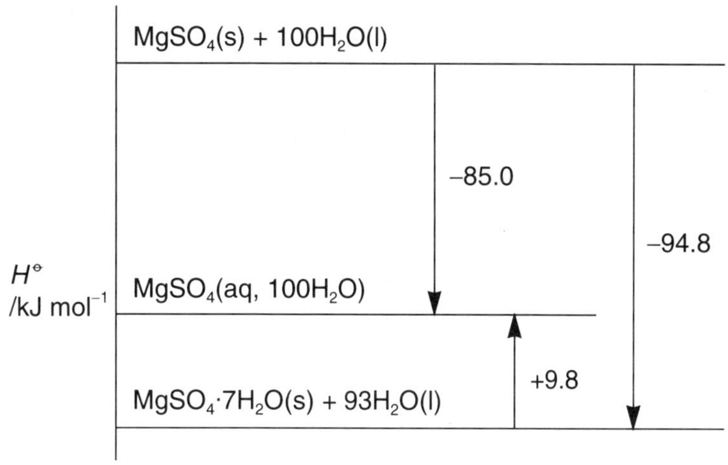

2. The maximum temperature change is much smaller and occurs more quickly.

3. Error $= \dfrac{-94.8 - (-104.0)}{104.0} \times 100 = \mathbf{-8.8\%}$

In addition to the usual assumptions about zero heat transfer between system and surroundings, and the heat capacities of the solutions, impurities in the salts may also affect the results. The anhydrous salt may not be totally free of water, and the hydrated salt may have absorbed extra water.

EXPERIMENT 2.4 Determining enthalpies of combustion

This experiment is lengthy and requires good experimental technique, but is well worth doing if time permits. If available, show one of the two ILPAC video programmes entitled 'Using a Heat of Combustion Apparatus' beforehand – the first refers to the glass envelope/copper spiral apparatus illustrated in the students' book, while the second refers to an improved all-metal apparatus (supplied by Philip Harris), which is recommended although slightly more expensive. Alternatively, you may wish to use the home-made apparatus described in Appendix 2 of the *Nuffield Advanced Chemistry Teachers' Guide II* – used with care, this can give quite good results. Whichever apparatus is used, it is well worthwhile obtaining six or seven **small** spirit lamps – they can be purchased separately from Griffin. This will eliminate errors arising from spillage and cleaning, and will save a good deal of time.

Some difficulty may be experienced in keeping the lamps burning satisfactorily during the experiment, particularly if the water pressure is such that filter pumps do not work very well. It is worth cleaning scaled-up tap apertures which can cut down water flow considerably. Standing the lamps on **small** blocks improves air flow but it is advisable to check quickly that the lamp will burn properly in position under the calorimeter **before** the temperature and weight are recorded.

It is important that for each determination the calorimeter should be refilled (to the mark) with fresh water from a supply which has stood in the laboratory overnight. Successive samples of tap water can vary significantly in temperature and this can invalidate the calibration of the apparatus. It is also important for the calorimeter to be used at exactly the same height for each determination.

Results tend to be low largely due to incomplete combustion but careful work should give a fairly constant difference between values for successive alcohols.

Specimen results and calculation

Results Table 2.3

	1st run	2nd run	
Molar mass of propan-1-ol, M	60.1	60.1	g mol^{-1}
Initial mass of spirit lamp + alcohol, m_1	13.112	12.083	g
Final mass of spirit lamp + alcohol, m_2	12.083	11.034	g
Mass of alcohol burned, $m_1 - m_2$	1.029	1.049	g
Amount of alcohol burned, $n = (m_1 - m_2)/M$	0.0171	0.0175	mol
Initial temperature of calorimeter	21.7	21.4	°C
Final temperature of calorimeter	33.2	33.0	°C
Temperature change, ΔT	11.5	11.6	K
Heat released during the experiment, ΔH $= \Delta H_c^\ominus$[propan-1-ol] $\times$ amount burned $= -2017$ kJ mol^{-1} (given) $\times n$	34.5	35.3	kJ
Heat required for a rise in temperature of 1 K $= \dfrac{-2017 \text{ kJ mol}^{-1} \times n}{\Delta T}$ = calibration factor, C	–3.00	–3.04	kJ K^{-1}
Average value of C		–3.02	kJ K^{-1}

Results Table 2.4

	C_4H_9OH	$C_5H_{11}OH$	$C_6H_{13}OH$	$C_7H_{15}OH$	$C_8H_{17}OH$
Molar mass, M/g mol^{-1}	74.1	88.2	102.2	116.2	130.2
Initial mass of lamp/g	13.691	12.820	13.571	13.679	13.909
Final mass of lamp/g	12.642	12.023	12.668	12.794	13.175
Mass of alcohol burned/g	1.049	0.797	0.903	0.885	0.734
Amount burned, n/mol	0.0142	0.00904	0.00884	0.00762	0.00564
Initial temperature/°C	21.5	21.0	21.7	21.4	22.0
Final temperature/°C	33.5	30.0	33.1	32.2	32.1
Temperature change, ΔT/K	12.0	9.0	11.4	10.8	10.1
$\Delta H_c = \dfrac{C \times \Delta T}{n}$/kJ mol^{-1}	–2552	–3007	–3895	–4280	–5408

EXPERIMENT 2.5 A thermometric titration

This experiment is straightforward and should give results within 2 or 3 kJ mol^{-1} of the data book values. Unfortunately, experimental error is sometimes more significant than the difference between the two values, but usually both results are low by about the same amount for the usual reasons – heat loss and assumptions about the heat capacities of the solutions.

The concentration of the sodium hydroxide solution must be precisely known – it is worth purchasing some ampoules of concentrated solution for making it up. It may be worth checking the concentrations of the acids to make sure that the equivalence point is **not** precisely at a multiple of 5 cm^3 of acid. This will ensure that students extrapolate their graphs properly.

Specimen results and calculation

Results Table 2.5
Titration of hydrochloric acid

Volume added/cm^3	0.0	5.0	10.0	15.0	20.0	25.0	30.0	35.0	40.0	45.0	50.0
Temperature/°C	22.2	24.4	26.4	28.4	30.1	31.1	30.4	29.9	29.2	28.8	28.2

Results Table 2.6
Titration of ethanoic acid

Volume added/cm^3	0.0	5.0	10.0	15.0	20.0	25.0	30.0	35.0	40.0	45.0	50.0
Temperature/°C	21.0	22.6	24.3	25.9	27.2	28.7	28.6	28.0	27.5	27.0	26.5

1 and 2.

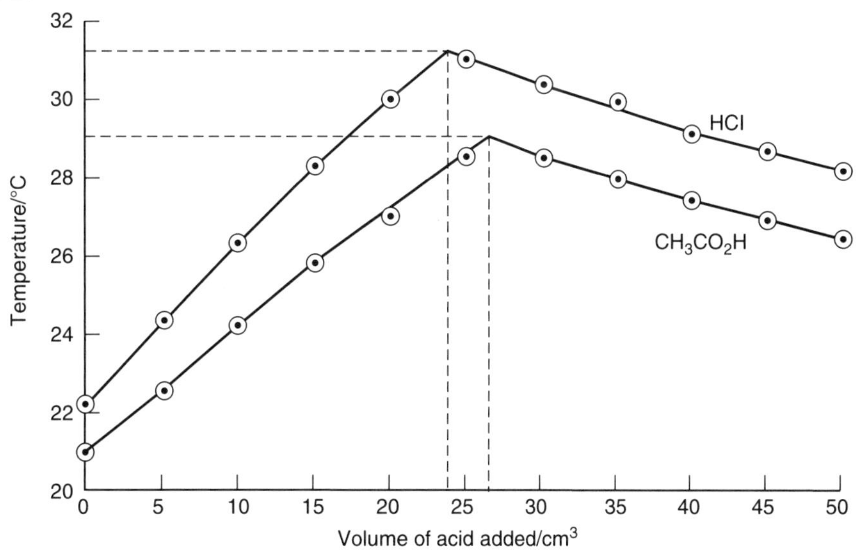

3. **a** $HCl(aq) + NaOH(aq) \rightarrow NaCl(aq) + H_2O(l)$
Amount of HCl = amount of NaOH

$$c \times \frac{23.5}{1000}\ dm^3 = 1.00\ mol\ dm^{-3} \times \frac{50.0}{1000}\ dm^3$$

$$\therefore c = 1.00\ mol\ dm^{-3} \times \frac{50.0}{23.5} = \mathbf{2.13\ mol\ dm^{-3}}$$

b $CH_3CO_2H(aq) + NaOH(aq) \rightarrow CH_3CO_2Na(aq) + H_2O(l)$
Amount of CH_3CO_2H = amount of NaOH

$$c \times \frac{26.5}{1000}\ dm^3 = 1.00\ mol\ dm^{-3} \times \frac{50.0}{1000}\ dm^3$$

$$\therefore c = 1.00\ mol\ dm^{-3} \times \frac{50.0}{26.5} = \mathbf{1.89\ mol\ dm^{-3}}$$

4 and 5. **a** Volume of mixture when reaction is complete = (50.0 + 23.5) cm^3 = 73.5 cm^3
$\Delta T = (31.3 - 22.2)\ K = 9.1\ K$
$\Delta H = -mc_p\Delta T = -0.0735\ kg \times 4.18\ kJ\ kg^{-1}\ K^{-1} \times 9.1\ K = -2.80\ kJ$
Amount of NaOH used = cV = $1.00\ mol\ dm^{-3} \times 0.0500\ dm^3 = 0.0500\ mol$

Scaling up to 1 mol, $\Delta H = -2.80\ kJ \times \frac{1}{0.0500} = -56.0\ kJ$

$\therefore HCl(aq) + NaOH(aq) \rightarrow NaCl(aq) + H_2O(l);\ \Delta H^\circ = \mathbf{-56.0\ kJ\ mol^{-1}}$

b Volume of mixture when reaction is complete = (50.0 + 26.5) cm^3 = 76.5 cm^3

$\Delta T = (29.1 - 21.0)$ K = 8.1 K

$\Delta H = -mc_p\Delta T = -0.0765$ kg × 4.18 kJ kg^{-1} K^{-1} × 8.1 K = −2.59 kJ

$$\text{Scaling up to 1 mol, } \Delta H = -2.59 \text{ kJ} \times \frac{1}{0.0500} = -51.8 \text{ kJ}$$

∴ $CH_3CO_2H(aq) + NaOH(aq) \rightarrow CH_3CO_2Na(aq) + H_2O(l)$; ΔH°= **−51.8 kJ mol^{-1}**

Answers to questions

1. The enthalpy change of neutralisation for completely ionised acids and bases is constant because the reaction is the same in every case.

$$H^+(aq) + OH^-(aq) \rightarrow H_2O(l)$$

2. Some heat is lost from the reaction mixture. The specific heat capacity of the mixture is not precisely 4.18 kJ kg^{-1} K^{-1}.
3. The enthalpy change of neutralisation for incompletely ionised acids and/or bases is less negative because some energy is required to complete the ionisation before the reaction between the hydrogen ion and hydroxide ion can occur.

EXPERIMENT 2.6 Using Hess' law (2)

The experimental procedure is straightforward but it takes longer than Experiment 2.3 and the calculation is more difficult – we suggest that weaker students do not attempt it. It may be worth purchasing a small amount of anhydrous copper sulphate each year since it is not too easy to dehydrate the crystals properly (see Technician's Sheet). Poor purity of the anhydrous salt, and its very slow rate of dissolving, could account for a smaller temperature rise than the 6.5 K needed to obtain a good result. Although the temperature change on dissolving the hydrate is much smaller (−1.2 K), students usually obtain a better result here.

Specimen results and calculations

Results Table 2.7

Mass of cold water in vacuum flask	50.0 g	50.0 g
Mass of warm water added	50.0 g	50.0 g
Initial temperature of flask and cold water	23.1 °C	23.5 °C
Initial temperature of warm water	41.3 °C	45.2 °C
Final temperature of flask and mixture	31.0 °C	33.0 °C

$$\begin{bmatrix}\text{Change in heat}\\ \text{energy of flask}\end{bmatrix} + \begin{bmatrix}\text{Change in heat}\\ \text{energy of cold water}\end{bmatrix} + \begin{bmatrix}\text{Change in heat energy}\\ \text{of warm water}\end{bmatrix} = 0$$

1) $[C \times (31.0 - 23.1)\text{K}] + [0.0500 \text{ kg} \times 4.18 \text{ kJ kg}^{-1} \text{ K}^{-1} \times (31.0 - 23.1) \text{ K}]$
$+ [0.0500 \text{ kg} \times 4.18 \text{ kJ kg}^{-1} \text{ K}^{-1} \times (41.3 - 31.0) \text{ K}] = 0$

$(C \times 7.9 \text{ K}) + (1.65 \text{ kJ}) + (-2.15 \text{ kJ}) = 0$

$$\therefore C = \frac{(2.15 - 1.65) \text{ kJ}}{7.9 \text{ K}} = 0.063 \text{ kJ K}^{-1}$$

2) $(C \times 9.5 \text{ K}) + (1.99 \text{ kJ}) + (-2.55 \text{ kJ}) = 0$

$$\therefore C = \frac{(2.55 - 1.99) \text{ kJ}}{9.5 \text{ K}} = 0.059 \text{ kJ K}^{-1}$$

Mean value of C = **0.061 kJ K^{-1}**

Results Table 2.8

Mass of anhydrous copper(II) sulphate	3.99 g
Mass of water	45.0 g
Initial temperature of vacuum flask and water	22.5 °C
Maximum temperature of vacuum flask and solution	27.5 °C

$$\begin{bmatrix}\text{Change in heat}\\ \text{energy of flask}\end{bmatrix}+\begin{bmatrix}\text{Change in heat}\\ \text{energy of contents}\end{bmatrix}+\begin{bmatrix}\text{Enthalpy change}\\ \text{of solution}\end{bmatrix}=0$$

$(0.061\ \text{kJ K}^{-1} \times 5.0\ \text{K}) + (0.0450\ \text{kg} \times 4.18\ \text{kJ kg}^{-1}\ \text{K}^{-1} \times 5.0\ \text{K}) + \Delta H = 0$
$0.305\ \text{kJ} + 0.941\ \text{kJ} + \Delta H = 0 \qquad \therefore\ \Delta H = -1.25\ \text{kJ}$

$$\text{Scaling up to 1 mol, } \Delta H = -1.25\ \text{kJ} \times \frac{1}{0.025} = -50\ \text{kJ}$$

$\therefore\ CuSO_4(s) + 100H_2O(l) \rightarrow CuSO_4(aq,\ 100H_2O);\ \Delta H^{\circ} = \mathbf{-50\ kJ\ mol^{-1}}$

Results Table 2.9

Mass of anhydrous copper(II) sulphate-5-water (0.0250 mol)	6.24 g
Mass of water added (0.025 × 95 mol)	42.75 g
Initial temperature of vacuum flask and water	23.0 °C
Maximum temperature of vacuum flask and solution	21.8 °C

$$\begin{bmatrix}\text{Change in heat}\\ \text{energy of flask}\end{bmatrix}+\begin{bmatrix}\text{Change in heat}\\ \text{energy of contents}\end{bmatrix}+\begin{bmatrix}\text{Enthalpy change}\\ \text{of solution}\end{bmatrix}=0$$

$0.061\ \text{kJ K}^{-1} \times (-1.2\ \text{K}) + (0.0450\ \text{kg} \times 4.18\ \text{kJ kg}^{-1}\ \text{K}^{-1} \times -1.2\ \text{K}) + \Delta H = 0$
(Note that the total mass of water in the contents is the same in both experiments.)
$-0.0732\ \text{kJ} + (-0.226\ \text{kJ}) + \Delta H = 0 \qquad \therefore\ \Delta H = +0.299\ \text{kJ}$

$$\text{Scaling up to 1 mol, } \Delta H = +0.299\ \text{kJ} \times \frac{1}{0.025} = +12.0\ \text{kJ}$$

$\therefore\ CuSO_4{\cdot}5H_2O(s) + 95H_2O(l) \rightarrow CuSO_4(aq,\ 100H_2O);\ \Delta H^{\circ} = \mathbf{+12.0\ kJ\ mol^{-1}}$

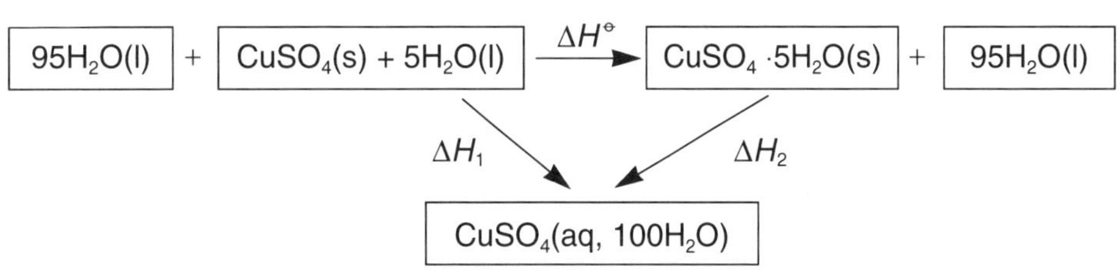

$\Delta H^{\circ} = \Delta H_1 - \Delta H_2 = -50\ \text{kJ mol}^{-1} - 11.5\ \text{kJ mol}^{-1} = \mathbf{-62\ kJ\ mol^{-1}}$

Answers to questions

1. The reaction would be extremely slow (even with excess water as in the experiment the students have done, it was quite slow).
2. Replacing lids and stoppers prevents moisture being absorbed from the air (especially important for anhydrous copper sulphate). Also, stoppers left off can be exchanged by mistake causing contamination of the contents when eventually replaced.
3. It is impossible to transfer **all** the weighed quantity from a beaker to the flask.

EXPERIMENT 3.1 Making models of two metallic structures

Any size of expanded polystyrene sphere can be used, but the larger the better for ease of handling.

Blu-tak seems to work rather better than Buddies.

Students should be encouraged to use adhesive on **all** the contacts mentioned in the instructions and to press spheres as close together as possible. In this way remarkably strong models can be assembled.

EXPERIMENT 3.2 Recognising ionic, covalent and metallic structures

Students should be discouraged from guessing the **identity** of the substances – they should be thinking about the relationships between **properties** and **structure**.

It might be worth disguising the white solids by grinding them to fine powders and mixing with a **little** carbon so that they all look grey. Students should also be encouraged to **select** tests rather than carry out five tests on each of the seven substances, and to be critical about the interpretation of their results.

Specimen results

Results Table 3.1

	A	B	C	D	E	F	G
Appearance	white powder	black powder	black powder	white powder	white powder	grey powder	white powder
Melting point (approx.)/°C	> 800	> 800	> 800	100	700	> 800	> 800
Solubility in water	insoluble	insoluble	insoluble	soluble	soluble	insoluble	insoluble
Conductivity of solution	–	–	–	none	good	–	–
Conductivity of solid	none	none	good	none	none	good	none
Action of dilute HCl	none	dissolves, green solution No H_2	none	dissolves, no evident reaction	dissolves, no evident reaction	dissolves, hydrogen evolved	none
Structure	ionic or covalent giant	ionic or covalent giant	metal or covalent giant	molecular crystal	ionic	metal	ionic or covalent giant

Notes on results

1. The heavy horizontal line in each column implies that students need not have done the tests below the line.
2. Students were not asked to identify the actual substances. However, for interest we list them:

   A = barium sulphate, $BaSO_4$; B = copper(II) oxide, CuO; C = graphite, C; D = glucose, $C_6H_{12}O_6$; E = potassium bromide, KBr; F = zinc, Zn; G = silica, SiO_2.
3. The students' general chemical knowledge probably enabled them to guess the identity of B and C, and hence to infer the structures. The tests alone, however, are not entirely conclusive.

Answers to questions

1. The use of higher temperatures and/or more sophisticated conductivity apparatus might enable the solids with high melting points to be classified as ionic or giant covalent structures.
2. It may not be easy to get a powdered metal to conduct electricity because the particles may not make good contact with each other, especially if there is an oxide film on the surfaces.

EXPERIMENT 3.3 Testing liquids for polarity

To minimise inhalation of vapours, corked burettes containing the liquids should be supplied for the students and the experiment should be done in a fume cupboard (with corks removed) if possible. Some of the liquids listed in the specimen results have been omitted from the students' instructions because of concerns about toxicity and ozone depletion.

It is difficult to standardise the procedure but the range of deflections should be broadly in line with the dipole moments of the liquids.

Specimen results

Results Table 3.2

Compound	Formula	Structural formula	Deflection
Cyclohexane	C_6H_{12}	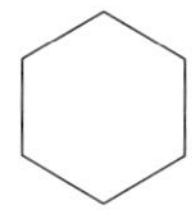	0
Cyclohexene	C_6H_{10}	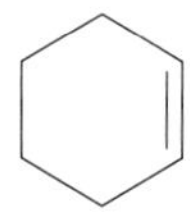	0 – 1
Hexane	C_6H_{14}	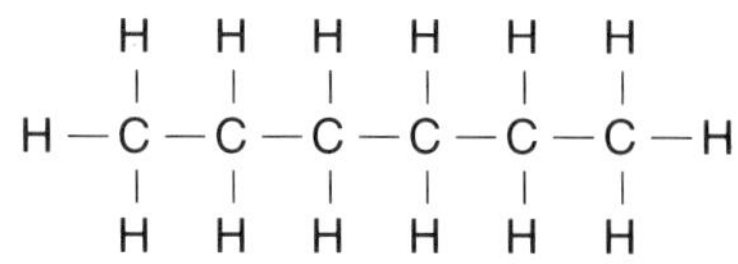	0
Hexan-1-ol	$C_6H_{13}OH$	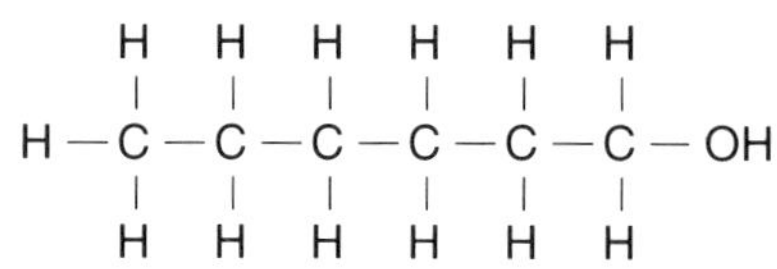	3
Methyl ethanoate	$CH_3CO_2CH_3$	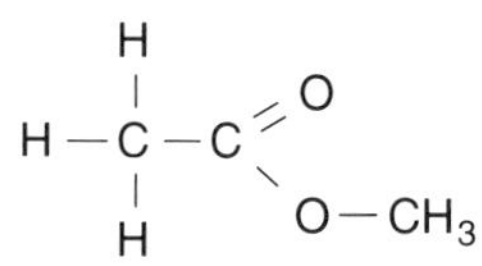	2
Propanone	CH_3COCH_3	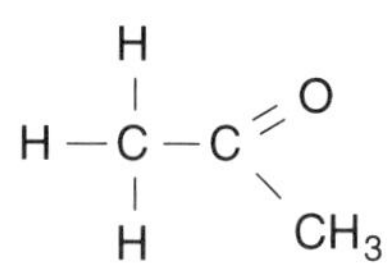	2
Tetrachloromethane	CCl_4	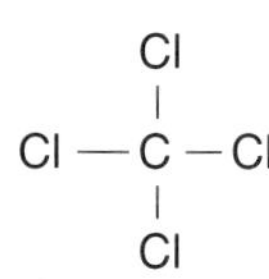	0
Trichloromethane	$CHCl_3$	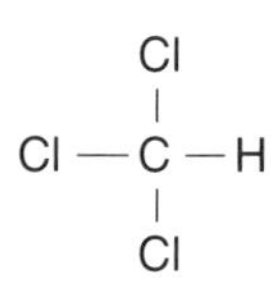	3
Water	H_2O	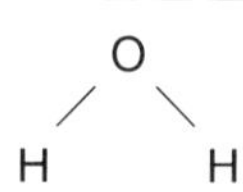	3

Answers to questions

1. The charged rod attracts that part of each polar molecule which has an opposite charge and repels that part with the same charge. This causes the molecules to align themselves along the lines of force and to move towards the charged rod.

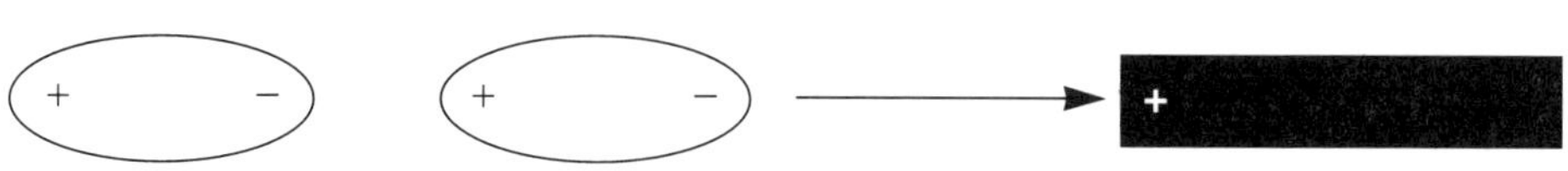

2. The molecules align themselves in the opposite sense but the observed movement is still in the same direction as before.

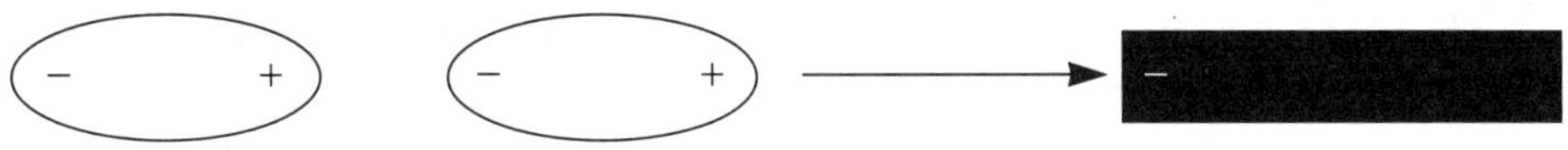

3. The approximate order of decreasing polarity is:

   water, trichloromethane, hexan-1-ol, methyl ethanoate, propanone, cyclohexene, followed by the three non-polar liquids, cyclohexane, hexane and tetrachloromethane.

   This order reflects both the relative polarity of bonds in the molecule and the degree of symmetry.

   a Both trichloromethane and tetrachloromethane contain highly polar C—Cl bonds, but in the latter case they are arranged symmetrically so that their effects cancel, making the molecule as a whole non-polar. Trichloromethane is strongly polar.

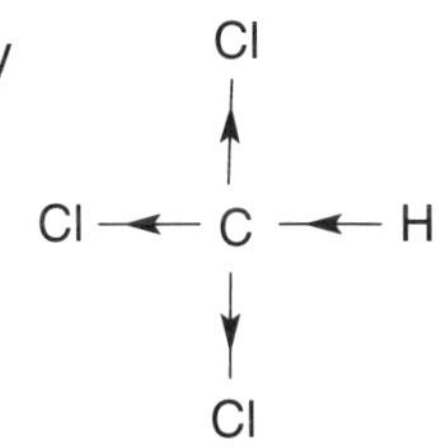

   b The cyclohexane molecule is symmetrical and is therefore non-polar. Cyclohexene is non-symmetrical; the net effect of the slightly polar C—H bonds makes the molecule slightly polar.

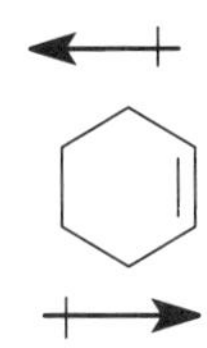

4. The denser the liquid, the smaller the deflection for the same attractive force. Variations in deflection might be partly due to variations in density.

EXPERIMENT 3.4 The effect of hydrogen bonding on liquid flow

It is worth keeping sealed tubes of these four liquids permanently in store; no further preparation is needed and the experiment takes only a few minutes.

Specimen results

Results Table 3.3

Liquid	Formula	Time/s
Propan-1-ol	$CH_3CH_2CH_2OH$	13
Propane-1,2-diol	$CH_3CH(OH)CH_2OH$	31
Propane-1,2,3-triol	$CH_2(OH)CH(OH)CH_2OH$	800
Propane-1,2,3-triyl triethanoate	$CH_2(OCOCH_3)CH(OCOCH_3)CH_2OCOCH_3$	17

Answers to questions

1. Each alcohol molecule forms hydrogen bonds with its neighbours through hydroxyl groups (—O—H·····O$<$). As the number of hydroxyl groups increases, so does the extent of hydrogen bonding in the liquid; this results in a great increase in viscosity.
2. Propane-1,2,3-triyl triethanoate has the stronger intermolecular van der Waals forces because of the greater number of electrons, but it cannot form hydrogen bonds because there are no hydrogen atoms with a sufficient positive charge. Propane-1,2,3-triol has extensive hydrogen bonding and this outweighs the effect of the weaker van der Waals forces.

EXPERIMENT 3.5 Variation of boiling point with composition of a mixture

This experiment should, ideally, be done in a fume cupboard. Students should be closely supervised to see that they handle flammable liquids with care and have minimum exposure to vapours. We suggest that each student, or pair, works with only one mixture in order to save time and materials.

Residues should be collected separately and disposed of as indicated in the Technician's Sheet for Experiment 3.3. Note that the trichloromethane/methyl ethanoate system should not now be used because of concerns about the toxicity of trichloromethane and its role as an ozone depleter. It has been suggested that mixtures of methanoic acid and butanone show a clear negative deviation, but the interpretation is more difficult. Therefore, we have retained specimen results for the trichloromethane/methyl ethanoate system for discussion purposes.

Suggested mark scheme

Total marks for manipulation = 16, as follows:

Assembly of apparatus		Use of apparatus	
Correct choice of components	(1)	Measurement of volumes	(2)
Lubrication of joints	(2)	Adjustment of heating	(2)
Correct support of assembly	(2)	Cooling between additions	(1)
Thermometer position (immersed, not touching wall)	(2)	Awareness of flammability	(2)
		Use of safety spectacles/goggles	(1)
Air gap	(1)		

Total marks	16–14	13–11	10–8	7–5	4–2	1–0
Scaled mark	5	4	3	2	1	0

Specimen results

Results Table 3.4

Volume/cm^3		% composition	Boiling point/°C		
A	B	A (by volume)	System 1	System 2	System 3
10	0	100	77	97.4	61.2
10	2	83.3	70	94.8	64.0
10	4	71.4	67.2	93.0	65.2
10	6	62.5	66	91.6	64.8
10	8	55.6	65.2	90.8	64.6
10	10	50	64.8	90.0	64.4
0	10	0	81	82.6	58.0
2	10	16.7	65.4	85.0	60.0
4	10	28.6	64.6	86.6	62.0
6	10	37.5	64.8	87.8	64.0
8	10	44.4	65.0	88.8	64.2

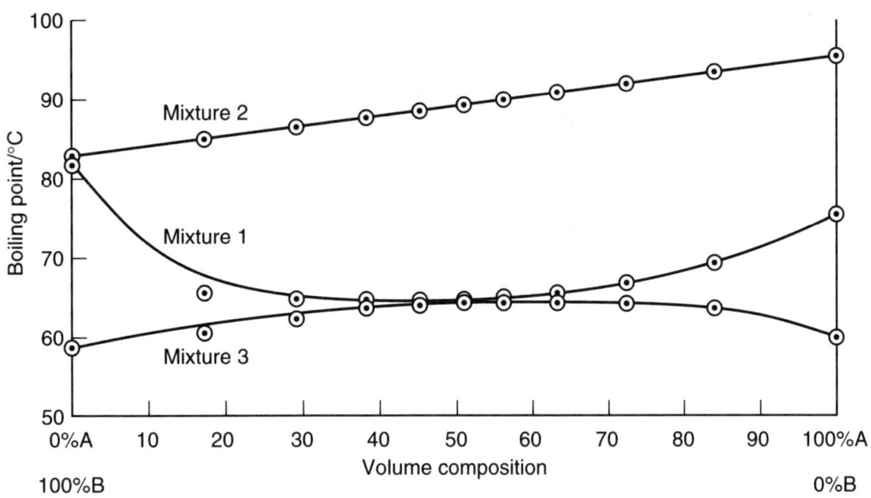

Answers to questions

1.

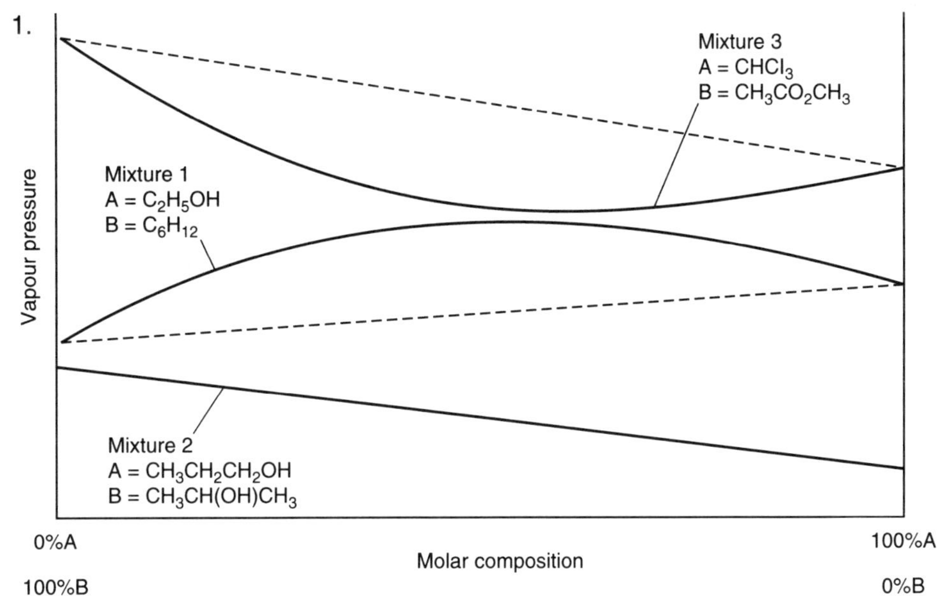

2. Mixture 1 (ethanol and cyclohexane) shows a positive deviation.
   Mixture 2 (propan-1-ol and propan-2-ol) is virtually ideal.
   Mixture 3 (trichloromethane and methyl ethanoate) shows a negative deviation from ideality.

EXPERIMENT 3.6 Measuring the temperature change on forming solutions

This is a very simple experiment and takes only a few minutes. We have omitted the calculation of values of ΔH_{mix} so that students focus on qualitative aspects of the experiment. However, we suggest that you remind students of relevant calculations, since such calculations could be required in this context.

See the note on Experiment 3.5 regarding disposal of residues and the use of the trichloromethane/methyl ethanoate system.

Specimen results
Results Table 3.5

Mixture	T_A/°C	T_B/°C	T_1/°C	T_2/°C	ΔT/°C
1. Ethanol and cyclohexane	24.0	24.0	24.0	20.5	–3.5
2. Propan-1-ol and propan-2-ol	24.0	24.5	24.5	24.5	0
3. Trichloromethane and methyl ethanoate	23.0	23.0	23.0	32.0	+9.0

Answers to questions

1. Mixture 1: ΔT is negative, $\therefore \Delta H^\ominus_{mix}$ is positive.
   Mixture 2: ΔT is zero, $\therefore \Delta H^\ominus_{mix}$ is zero.
   Mixture 3: ΔT is positive, $\therefore \Delta H^\ominus_{mix}$ is negative.
2. Mixture 2, propan-1-ol and propan-2-ol, is virtually ideal. There is no energy change on mixing, which shows that the A—B bonds formed are of the same strength as the A—A and B—B bonds which they replace.
3. The temperature increases when trichloromethane and methyl ethanoate are mixed because more energy is released in making A—B bonds than is required to break the A—A and B—B bonds which they replace.
4. Since molecules of trichloromethane and methyl ethanoate are more strongly bound together in the mixture than molecules in the pure liquids, their tendency to escape into the vapour phase is reduced. This causes a reduction in vapour pressure compared with an ideal mixture. The effect is great enough to cause a minimum in the vapour pressure/composition curve.
5. The positive value of $\Delta H^\ominus_{mix}$ for the ethanol–cyclohexane system shows that A—B bonds are weaker than the A—A and B—B bonds. This gives a higher vapour pressure than ideal and the effect is great enough to give a maximum in the vapour pressure/composition curve.

EXPERIMENT 4.1 Reaction between sodium peroxide and water

Sodium peroxide must be reasonably fresh – it absorbs carbon dioxide from the air, turning from pale yellow to white in the process. Teachers may be more familiar with a test for hydrogen peroxide using ethoxyethane (diethyl ether), but 2-methylbutan-1-ol is safer and gives the same result provided the procedure is carefully followed. Ethyl ethanoate may be used instead of 2-methylbutan-1-ol.

Specimen results
Results Table 4.1

Experiment	Observation	Conclusion
Sodium peroxide added to water plus universal indicator	Colour of gas: None Effect on glowing splint: Relights pH of water: 12 or higher	The gas is oxygen
Sodium peroxide plus water, added to acidified dichromate solution plus 2-methylbutan-1-ol	Colour of organic (top) layer is: Blue Colour of aqueous (bottom) layer is: Very pale green	The colour of the organic layer shows that a peroxide has been formed

Answers to questions

1. **a** $Na_2O_2(s) + 2H_2O(l) \rightarrow 2NaOH(aq) + H_2O_2(aq)$
   b $2H_2O_2(aq) \rightarrow 2H_2O(l) + O_2(g)$
   c $2Na_2O_2(s) + 2H_2O(l) \rightarrow 4NaOH(aq) + O_2(g)$
2. **a** $BaO_2(s) + 2H_2O(l) \rightarrow Ba(OH)_2(aq) + H_2O_2(aq)$
   b i) $BaO_2(s) + H_2SO_4(aq) \rightarrow BaSO_4(s) + H_2O_2(l)$

ii) It produces barium sulphate, which is insoluble and can be separated easily from the hydrogen peroxide by filtration.

3. **a** $Na_2O(s) + CO_2(g) \rightarrow Na_2CO_3(s)$
   b $2Na_2O_2(s) + 2CO_2(g) \rightarrow 2Na_2CO_3(s) + O_2(g)$
   c $4KO_2(s) + 2CO_2(g) \rightarrow 2K_2CO_3(s) + 3O_2(g)$

EXPERIMENT 4.2 Heating the nitrates and carbonates of s-block elements

This is a very simple experiment and should not be allowed to take up too much laboratory time, particularly if students are familiar with these reactions from pre-A-level work. Encourage students to see the importance of following the same procedure as precisely as possible for successive carbonates or nitrates. Decomposition times may vary widely from our specimen results but should show the same trends. We have found that potassium and sodium carbonates seem to decompose more readily than textbooks suggest.

Specimen results

Results Table 4.2
Effect of heat on the s-block carbonates

Carbonate	Time to detect CO_2	Observations
Li_2CO_3	35 s	Lime-water very cloudy
Na_2CO_3	50 s	Slight cloudiness
K_2CO_3	70 s	Slight cloudiness
Rb_2CO_3	not available	–
Cs_2CO_3	65 s	Slight cloudiness
$MgCO_3$	20 s	Lime-water very cloudy
$CaCO_3$	40 s	Lime-water very cloudy
$SrCO_3$	not available	–
$BaCO_3$	no CO_2 detected	No change observed

Results Table 4.3
Effect of heat on the s-block nitrates

Nitrate	Time to detect O_2 or NO_2	Observations	Effect of adding dilute HCl to cold residue
$LiNO_3$	NO_2 – 35 s	Liquid formed immediately – steamy gas evolved – resolidified after 20 s	No reaction
$NaNO_3$	O_2 – 40 s	Liquid formed after 20 s – no steamy gas – bubbles formed in liquid	Brown fumes evolved
KNO_3	O_2 – 50 s	As for $NaNO_3$	Brown fumes evolved
$RbNO_3$	not available	–	
$CsNO_3$	O_2 – 60 s	As for $NaNO_3$	Brown fumes evolved
$Mg(NO_3)_2$	NO_2 – 25 s	Liquid formed immediately – steamy gas evolved – resolidified after about 20 s	No reaction
$Ca(NO_3)_2$	NO_2 – 30 s	As for Mg	No reaction
$Sr(NO_3)_2$	NO_2 – 32 s	White powder crackled – no melting	No reaction
$Ba(NO_3)_2$	NO_2 – 42 s	White powder melted after 30 s	No reaction

Answers to questions

1. Some of these nitrates are hydrated and therefore dissolve in their own water of crystallisation. The water eventually boils off leaving a white solid.
2. Lithium nitrate and lithium carbonate.
3. Thermal stability of both the nitrates and carbonates increases as each group is descended.
4. Group I nitrates and carbonates are more stable.
5. **a** $4LiNO_3(s) \rightarrow 2Li_2O(s) + 4NO_2(g) + O_2(g)$
   $Li_2CO_3(s) \rightarrow Li_2O(s) + CO_2(g)$
   b $2KNO_3(s) \rightarrow 2KNO_2(l) + O_2(g)$. The liquid product is potassium nitr**ite**.
   K_2CO_3 is usually reckoned to be thermally stable at Bunsen burner temperatures, although traces of CO_2 may be detected.
   c $2Mg(NO_3)_2(s) \rightarrow 2MgO(s) + 4NO_2(g) + O_2(s)$
   $MgCO_3(s) \rightarrow MgO(s) + CO_2(g)$
6. All Group I nitrates (except $LiNO_3$) decompose to the nitrite, which produces NO_2 (amongst other nitrogen compounds) if warmed with dilute acids. This is a useful test for the presence of nitrite ions.

EXPERIMENT 4.3 The solubility of some salts of Group II elements

This experiment works well provided the solutions are of the right concentrations, and the trends in solubility should be readily apparent. The sodium hydroxide solution should be fresh since contamination with carbonate can cause unwanted precipitates.

Specimen results

Results Table 4.4

Cation solution	Number of drops of anion solution added to give a precipitate			
	OH^-	SO_4^{2-}	SO_3^{2-}	CO_3^{2-}
Mg^{2+}	5	40+	40+	25
Ca^{2+}	20	40+	10	8
Sr^{2+}	30 (slight)	8	2	4
Ba^{2+}	40+	1	1	2

Answers to questions

Students' results will probably differ from the specimen results above, in the numbers of drops used, but the trends should be fairly clear.

a The solubility of the hydroxides increases down Group II.
b The solubility of the sulphates decreases down Group II.
c The solubility of the sulphites decreases down Group II.
d The solubility of the carbonates decreases down Group II.

EXPERIMENT 4.4 The solubility of the halogens in organic solvents

Marked colour changes indicate that bromine and iodine are more soluble in these organic solvents than in water. Chlorine is also more soluble in organic solvents than in water, but this cannot be deduced directly from the experiments because colour changes are scarcely noticeable.

We recommend that chlorinated hydrocarbons are **not** used as solvents in this experiment because of concerns about health hazards and environmental damage (ozone depletion).

Specimen results

Results Table 4.5

	Nature of each layer	Chlorine water	Bromine water	Iodine solution
Colour of aqueous solution		faint green	orange–brown	reddish brown
Colour of each layer after shaking with Volasil 244	Upper layer aqueous	no colour	red–brown	violet
	Lower layer organic	no colour	yellow	light brown
Colour of each layer after shaking with ethoxyethane	Upper layer organic	no colour	orange–brown	light brown
	Lower layer aqueous	no colour	yellow	yellow
Colour of each layer after shaking with cyclohexane	Upper layer organic	no colour	orange–brown	violet
	Lower layer aqueous	no colour	yellow	light brown

Answers to questions

1. The halogens are more soluble in organic solvents than in water because if dilute aqueous solutions of the coloured halogens are mixed with organic solvents, most of the colour is transferred to the organic solvent.
2. Iodine has a significantly different colour in the organic layer than in the aqueous layer. Iodine in hexane is purple (or violet). Iodine in dichloromethane is also purple when dilute, but red when more concentrated.
3. To distinguish between a dilute solution of iodine and a more concentrated solution of bromine, add a few cm^3 of hexane to each solution and shake. The iodine appears purple in hexane while the bromine appears brown.

EXPERIMENT 4.5 The action of dilute alkali on the halogens

This is a very straightforward experiment and should take only a few minutes to perform. However, make sure your students use their books to answer the questions properly.

Specimen results

Results Table 4.6

Aqueous halogen	Original colour	Colour after adding NaOH(aq)	Colour after adding H_2SO_4(aq)
Bromine water	orange-brown	very pale yellow	orange-brown
Iodine solution	reddish-brown	very pale yellow	reddish-brown

Answers to questions

1. $3Br_2(aq) + 6OH^-(aq) \rightarrow 5Br^-(aq) + BrO_3^-(aq) + 3H_2O(l)$

   $3I_2(s) + 6OH^-(aq) \rightarrow 5I^-(aq) + IO_3^-(aq) + 3H_2O(l)$
2. Both of the above reactions are disproportionation reactions.
3. Both reactions are reversible because the halogen colour returns in both cases when the colourless halogen–alkali mixtures are acidified.
4. Chlorine reacts with cold alkali to form chlorate(I) ion, ClO^-.

   $Cl_2(aq) + 2OH^-(aq) \rightarrow Cl^-(aq) + ClO^-(aq) + H_2O(l)$

   $ClO^-(aq)$ is stable at room temperature whereas BrO^- and IO^- are not.

EXPERIMENT 4.6 Halogen–halide reactions in aqueous solution

If your students have done this experiment in their pre-A-level course, you may feel they need not repeat it here. However, make sure they understand the use of a suitable solvent for identification, especially to distinguish between bromine and iodine.

Cyclohexane may be used instead of Volasil. You may also wish to introduce the use of starch solution as a test for iodine.

Specimen results

Results Table 4.7

			Chlorine water	Bromine water	Iodine solution
1.	Initial colour		almost none	orange	brown
2.	Colour after shaking with KI		brown	brown	
	Colour of each layer after shaking with Volasil 244	Upper	violet	violet	
		Lower	yellow	yellow	
	Conclusion		I^- oxidised	I^- oxidised	
3.	Colour after shaking with KBr		orange		unchanged
	Colour of each layer after shaking with Volasil 244	Upper	red		violet
		Lower	almost none		yellow
	Conclusion		Br^- oxidised		no reaction
4.	Colour after shaking with KCl			same	same
	Colour of each layer after shaking with Volasil 244	Upper		red	violet
		Lower		almost none	yellow
	Conclusion			no reaction	no reaction

Answers to questions

1. **a** $I_2(aq)$ does not oxidise $Cl^-(aq)$ or $Br^-(aq)$.
   b $Br_2(aq)$ oxidises $I^-(aq)$ but not $Cl^-(aq)$.
   c $Cl_2(aq)$ oxidises both $Br^-(aq)$ and $I^-(aq)$.
2. $Cl_2(aq) + 2Br^-(aq) \rightarrow 2Cl^-(aq) + Br_2(aq)$
   $Cl_2(aq) + 2I^-(aq) \rightarrow 2Cl^-(aq) + I_2(aq)$
   $Br_2(aq) + 2I^-(aq) \rightarrow 2Br^-(aq) + I_2(aq)$

EXPERIMENT 4.7 Reactions of solid halides

Make sure your students take care in using concentrated sulphuric acid and know how to deal with minor spillages. Orthophosphoric(V) acid should also be treated with care. Some success may be obtained using 'syrupy' phosphoric acid, containing about 10% of water, but 100% phosphoric acid, which is a solid, is better. Syrupy phosphoric(V) acid can be dehydrated by cautiously adding phosphorus(V) oxide, P_4O_{10}.

Specimen results

Results Table 4.8

Test	**Chloride**	**Bromide**	**Iodide**
Action of conc. H_2SO_4 and MnO_2	Steamy fumes in cold. Pale green gas on warming. Choking smell	Steamy brown fumes in cold. More on warming. Choking smell	Steamy violet fumes in cold. More on warming. Bad egg smell
Suspected product	$Cl_2(HCl)$	$Br_2(HBr)$	$I_2(HI, H_2S)$
Confirmatory test	Blue litmus bleached. Starch/ iodide paper blue	Red colour when bubbled into hexane	Violet colour in hexane
Action of conc. H_2SO_4	Steamy fumes in cold. No green gas even on heating	Steamy fumes in cold. Brown gas on warming	Steamy violet fumes in cold. More on warming Bad egg smell.
Suspected product	HCl	$HBr(Br_2)$	$I_2(HI, H_2S)$
Confirmatory test	White smoke with ammonia. Blue litmus turns red but not bleached	White smoke with ammonia. Red colour in hexane	Violet colour in hexane
Action of H_3PO_4	Steamy fumes on warming	Steamy fumes on warming	Steamy fumes on warming
Suspected product	HCl	HBr	HI
Confirmatory test	White smoke with ammonia	White smoke with ammonia	White smoke with ammonia

Answers to questions

1. **a** $KCl(s) + H_2SO_4(l) \rightarrow KHSO_4(s) + HCl(g)$ (K_2SO_4 at high temperature only)

   b $2KCl(s) + H_3PO_4(l) \rightarrow K_2HPO_4(s) + 2HCl(g)$

   c $KBr(s) + H_2SO_4(l) \rightarrow HBr(g) + KHSO_4(s)$

   $2HBr(g) + H_2SO_4(l) \rightarrow SO_2(g) + 2H_2O(l) + Br_2(l)$

   d $2KBr(s) + H_3PO_4(l) \rightarrow K_2HPO_4(s) + 2HBr(g)$

   e $KI(s) + H_2SO_4(l) \rightarrow KHSO_4(s) + HI(g)$

   $2HI(g) + H_2SO_4(l) \rightarrow SO_2(g) + 2H_2O(l) + I_2(s)$ or

   $8HI(g) + H_2SO_4(l) \rightarrow H_2S(g) + 4I_2(s) + 4H_2O(l)$

   f $2KI(s) + H_3PO_4(l) \rightarrow K_2HPO_4(s) + 2HI(g)$

2. Concentrated H_2SO_4 alone is not a sufficiently strong oxidising agent to oxidise $HCl(g)$ to $Cl_2(g)$. This is due to the strong H—Cl bond. The combination of concentrated H_2SO_4 and MnO_2 can oxidise $HCl(g)$ to $Cl_2(g)$.
3. Concentrated H_2SO_4 alone is a sufficiently strong oxidising agent to oxidise HI to I_2. This is due to the weakness of the H—I bond. A more powerful oxidising agent such as concentrated H_2SO_4 and MnO_2 will merely produce similar results.

EXPERIMENT 4.8 Reactions of halides in solution

Your students may have done some of these test-tube experiments before, but they are probably worth repeating quickly. You may choose to introduce the formulae of complex ions, such as $[Ag(NH_3)_2]^+$, to explain the solubility of silver halides in ammonia solution, but a detailed study is best left until later.

Specimen results

Results Table 4.9

Test	Chloride	Bromide	Iodide
Action of $AgNO_3(aq)$	White ppt.	Pale yellow ppt.	Yellow ppt.
Effect of standing in **a** dark **b** light	 No change Darkens rapidly	 No change Darkens slightly	 No change No change
Action of $AgNO_3(aq)$ followed by dilute $HNO_3(aq)$	White ppt. insoluble in $HNO_3(aq)$	Pale yellow ppt. insoluble in $HNO_3(aq)$	Yellow ppt. insoluble in $HNO_3(aq)$
Action of $AgNO_3(aq)$ followed by $NH_3(aq)$	White ppt. soluble in $NH_3(aq)$	Pale yellow ppt. slightly soluble in $NH_3(aq)$	Yellow ppt. insoluble in $NH_3(aq)$
Action of $Pb(NO_3)_2(aq)$	White ppt.	White ppt.	Yellow ppt.
Action of $H_2O_2(aq)$ and dilute $H_2SO_4(aq)$	No change	No change	Brown colour. Violet in hexane

Answers to questions

1. **a** $Ag^+(aq) + Cl^-(aq) \rightarrow AgCl(s)$

   $Ag^+(aq) + Br^-(aq) \rightarrow AgBr(s)$

   $Ag^+(aq) + I^-(aq) \rightarrow AgI(s)$

   b $Pb^{2+}(aq) + 2Cl^-(aq) \rightarrow PbCl_2(s)$

   $Pb^{2+}(aq) + 2Br^-(aq) \rightarrow PbBr_2(s)$

   $Pb^{2+}(aq) + 2I^-(aq) \rightarrow PbI_2(s)$

2. **a** Add silver nitrate solution to each solution in turn, followed by dilute or concentrated ammonia solution. A white precipitate is obtained with the chloride ion, which is soluble in dilute ammonia solution. A pale yellow precipitate is obtained with the bromide ion, which is slightly soluble in dilute ammonia solution and soluble in concentrated ammonia solution.

   b Add lead ions to bromide and iodide solutions separately. A white precipitate is obtained with the Br^- ions and a yellow precipitate with the I^- ions.
3. **a** $2I^-(aq) + H_2O_2(aq) + 2H^+(aq) \rightarrow I_2(aq) + 2H_2O(l)$

   b H_2O_2 is not a sufficiently strong oxidising agent to oxidise $Cl^-(aq)$ and $Br^-(aq)$ ions.
4. The light converts some of the silver halide to small particles of metallic silver, which darkens the precipitates.

EXPERIMENT 4.9 Balancing a redox reaction

This experiment could be used for assessment of manipulative skills provided you are confident that your students have reached an adequate level of skill in titration but make sure your students know that accuracy will be assessed here. Brighter students might otherwise realise that a high degree of accuracy is not strictly necessary – a fairly approximate titre (within 0.5 cm^3 of the correct value) would still indicate the correct integer as the final result. You will have to do the experiment yourself to establish the 'correct' titre and this should be done immediately before or after the students to ensure reliability – you will not have time to observe students properly if you do it at the same time! Make sure the starch solution is freshly prepared so that a clear end-point can be observed.

Suggested mark scheme

Total marks for manipulation = 12, as follows:

Safety glasses worn throughout (1)
General tidiness of work area (1)
Correct use of measuring cylinder for KI (1)
Correct use of burette for KIO_3 (meniscus at eye-level) (Accuracy included below) (1)
Addition of starch at right time (1)
Visual determination of end-point (2)
Consistency of best two titres (±0.15 cm^3 = 2 marks; ±0.20 cm^3 = 1 mark) (2)
Agreement with correct titre
(±0.15 cm^3 = 3 marks; ±0.25 cm^3 = 2 marks; ±0.35 cm^3 = 1 mark). N.B. Better accuracy than this would be expected if the KIO_3 were pipetted (3)

Total marks	12–11	10–9	8–6	5–4	3–2	1–0
Scaled mark	5	4	3	2	1	0

Specimen results

Results Table 4.10

Solution in flask	iodide/iodate/acid mixture					25.0 cm^3
Solution in burette	sodium thiosulphate			0.10 mol dm^{-3}		
Indicator	starch					
		Trial	1	2	3	4
Burette readings	Final	30.4	30.30	30.20	30.40	–
	Initial	0.0	0.00	0.00	0.20	–
Volume used/cm^3		30.4	30.30	30.20	30.20	–
Mean titre/cm^3	30.20					

Calculations

1. $n = cV = 0.10\ mol\ dm^{-3} \times (30.2/1000)\ dm^3 = 30.2 \times 10^{-4}\ mol$
2. Amount of iodine atoms = amount of thiosulphate = 30.2×10^{-4} mol
3. $n = cV = 0.10\ mol\ dm^{-3} \times (5.0/1000)\ dm^3 = 5.0 \times 10^{-4}\ mol$
4. Amount of I from KI = total amount – amount from KIO_3
   $= (30.2 - 5.0) \times 10^{-4}$ mol
   $= 25.2 \times 10^{-4}$ mol
5. 5.0×10^{-4} mol of KIO_3 reacts with 25.2×10^{-4} mol of KI
   ∴ 1 mol of KIO_3 reacts with (25.2/5.0) mol of KI = 5 mol
   This is in agreement with the equation
   $IO_3^-(aq) + 6H^+(aq) + 5I^-(aq) \rightarrow 3I_2(aq) + 3H_2O(l)$

EXPERIMENT 4.10 Observation and deduction exercise 1

Most students need a lot of help when first performing experiments of this sort, in making observations, in interpreting them and in reporting **fully** what they see and deduce. We suggest that you spend as much time as possible discussing the experiments with individual students while they are performing them. Make sure your students realise that in an examination, many marks can be obtained for detailed and accurate reporting of observations, even when no firm conclusions are drawn from them.

If you feel that your students already have sufficient skill in this area, you may wish to use Experiment 4.10 for assessment of both observation and interpretation based on the specimen results tables below, which are not included in the student's book.

Suggested mark scheme

Total marks for observation and recording = 10, as follows:

a Decrepitation (or equivalent wording) (1)
Melting to clear liquid which solidifies on further heating (2)
Colourless gas relights glowing splint (1)
Solid formed as above melts on strong heating (1)
Liquid cools to colourless solid (1)

b i) creamy white precipitate, insoluble in dilute HNO_3 (2)
ii) yellow/brown solution (1)
iii) white precipitate (1)

Total marks	10–9	8–7	6–5	4–3	2–1	1–0
Scaled mark	5	4	3	2	1	0

Total marks for interpretation = 14, as follows:

a Oxygen evolved, from an oxysalt (2)

b i) Silver halide precipitated (1)
ii) Bromine or iodine displaced from a bromide or iodide (2)
iii) Lead bromide precipitated, **not** iodide (2)

Non-metal – bromine (1)
Substances/ions – bromate(V), BrO_3^-, and bromide, Br^- (4)
Oxidation numbers – +5 and –1 (2)

Total marks	14–12	11–9	10–7	6–5	4–3	2–0
Scaled mark	5	4	3	2	1	0

Specimen results

Results Table 4.11

Test	Observations	Inferences
a Heat approximately 0.1 g of F in a Pyrex tube, at first gently and then more strongly, until the change is complete. Cool and keep the residue. Test any gases evolved	White solid decrepitates on heating giving first a liquid, then a white solid. A colourless gas is given off which relights a glowing splint. On further heating, the white solid melts to a clear liquid which eventually turns orange. On cooling, the orange liquid becomes colourless and finally solidifies to a white solid	O_2 evolved. Orange solution suggests the formation of Br_2. F could be BrO_3^- or IO_3^-. The final residue is probably an alkali metal bromide or iodide
b Make an aqueous solution of the residue from **a** and carry out the following tests on portions:		
i) add aqueous silver nitrate followed by acid	A cream precipitate forms which is insoluble in dilute HNO_3	AgBr (or possibly AgI) precipitated from Br^- (or I^-) in residue
ii) add aqueous chlorine	A yellow solution is produced	Br_2 (or possibly I_2) displaced by Cl_2 from halide
iii) add aqueous lead(II) ethanoate (lead acetate)	A white precipitate is formed	Confirms Br^- not I^- in the residue (PbI_2 is yellow)

Answer to question

Results Table 4.12

Substance provided	Name of non-metal	Name and formula of substance/ion	Oxidation number
F	Bromine	Bromate(V), BrO_3^- Bromide, Br^-	+V (+5) –I (–1)

EXPERIMENT 4.11 Observation and deduction exercise 2

Please see notes for Experiment 4.10.

Specimen results

Results Table 4.13
Tests with unknown substance D

Method	Observations	Inferences
a A few crystals of D were heated in an ignition tube, gently at first and then more strongly	Some decrepitation at first. Solid eventually melted to a clear liquid which turned brown, then red, on further heating and gave off a little purple vapour. Cooled to a colourless solid	Purple vapour must be iodine, probably from thermal decomposition of potassium iodide
b The residue from **a** was cooled and one drop of concentrated H_2SO_4 was added. Then a few more drops were added	The mixture immediately darkened (almost black) and steamy fumes with a choking and 'bad egg' smell were released. A brownish liquid condensed on the tube just above the mixture	HI produced (displaced from iodide by non-volatile H_2SO_4). Some oxidation of HI to iodine by concentrated H_2SO_4, which was reduced to H_2S
c 10 drops of dilute H_2SO_4 were added to 5 drops of $KMnO_4$ solution followed by a few crystals of D. The mixture was then warmed	The solution immediately changed from purple to dark brown. No further change on warming	I^- oxidised to I_2 (brown in solution as I_3^-)

See page 38 for Results Table 4.14.

Results Table 4.15
Experiments to test inferences

Inference tested	Test and observations	Conclusion
Iodine produced by reaction of D and E in acid solution	About 1 cm^3 of Volasil 244 was added to mixture of D, E and acid (**e** in Results Table 4.14). The tube was shaken gently The upper layer of Volasil 244 turned purple	Iodine confirmed
D is potassium iodide	A few crystals of D were dissolved in distilled water and a few drops of dilute HNO_3 were added followed by $AgNO_3$ solution Yellow precipitate formed	Yellow precipitate of AgI confirms that D is potassium iodide

Final conclusion

D is potassium iodide, KI, and E is potassium iodate(V), KIO_3.

Results Table 4.14 Tests with unknown substance E

	Method	Observations	Inferences
a	A few crystals of E were heated in an ignition tube, gently at first and then more strongly	No change at first, but solid soon melted to colourless liquid and effervesced. Gas was colourless at first, then tinged with purple. Red liquid eventually formed on strong heating which cooled to a colourless solid	Purple vapour must be iodine but E is not the same as D. Perhaps D is produced from E by decomposition
	Gas was tested with damp litmus papers, blue and red	No colour change	Neutral gas (or CO_2) produced
	Gas was drawn into a teat pipette and bubbled through a little lime-water	No change	No CO_2 produced
	A glowing splint was lowered into the tube	The splint relit	O_2 produced, probably by decomposition of potassium iodate(V)
b	The residue from **a** was cooled and one drop of conc. H_2SO_4 was added. Then a few more drops were added	The mixture immediately turned almost black and colourless fumes with a choking 'bad egg' smell were released. A brownish liquid conden-sed on the tube above the mixture	HI produced (displaced from iodide by non-volatile H_2SO_4). Some oxidation of HI to iodine by concentrated H_2SO_4, which was reduced to H_2S. This suggests that the residue contains I^-, thus E is very likely to be KIO_3
c	10 drops of dilute H_2SO_4 were added to 5 drops of $KMnO_4$ solution followed by a few crystals of E. The mixture was then warmed	No change observed apart from dissolving of crystals	No redox reaction occurs. This is consistent with the identity suggested for E, i.e. KIO_3
d	A few crystals of D and E were dis-solved separately in about 2 cm^3 of distilled water in each of two tubes. The solutions were then mixed	D dissolved very easily, E less easily. Both solutions were colour-less and remained so on mixing (perhaps a slight yellow colouration)	No apparent reaction. This again is consistent with KI(aq) and KIO_3(aq) in neutral solution
e	Dilute H_2SO_4 was added dropwise to the mixed solu-tions from **d**	Solution turned dark brown with some blackish precipitate	I_2 formed by redox reac-tion between I^- and IO_3^- in presence of H^+

EXPERIMENT 4.12 Investigating the properties of Period 3 chlorides

In this experiment we have asked students to use anhydrous aluminium chloride and hydrated magnesium chloride. You may feel that this is an unfair comparison and decide to use anhydrous magnesium chloride which is available cheaply as a technical grade reagent. This solid gives a temperature rise of about 7 °C when added to water but the resulting solution is strongly alkaline. We have therefore decided that the technical grade is too impure for the purposes of this experiment. Hydrated aluminium chloride gives a very small temperature rise when added to water (about 1 °C) and the resulting solution is strongly acidic. This could be used in place of anhydrous aluminium chloride.

Silicon tetrachloride is readily hydrolysed and the product can seal a stopper firmly in place and allow pressure to build up in the bottle. Bottles containing $SiCl_4$ have been known to burst in storage, or on being opened. They should therefore be opened in the fume cupboard with considerable caution after being covered with a cloth. It is particularly dangerous to return unused chemical to the bottle. When disposing of organic residues ask your students to avoid pouring undissolved solid into the residues bottle.

Specimen results

Results Table 4.16

	$NaCl$	$MgCl_2$	$AlCl_3$	$SiCl_4$	PCl_3	S_2Cl_2
Appearance	Colourless solid	Colourless solid	Colourless solid	Colourless liquid	Colourless liquid	Yellow liquid
On mixing with water		(Hydrated)	(Anhydrous)			
Initial temp.	17 °C	17 °C	17 °C	17 °C	17 °C	17 °C
Final temp.	17 °C	19 °C	49 °C	19 °C	27 °C	20 °C
Does it dissolve?	Yes	Yes	Yes	Yes	Yes (eventually)	Yes
pH of solution	7	6	4 or less	Below 4	Below 4	Below 4
Other observations (if any)	None	None	Vigorous reaction. Acid gas – white fumes with NH_3	Acid gas – white fumes with NH_3. White particles in liquid	Acid gas – white fumes with NH_3	Cloudy solution formed
On mixing with cyclohexane						
Initial temp.	16 °C	16 °C	16 °C	16 °C	17 °C	17 °C
Final temp.	16 °C	16 °C	16 °C	16 °C	17 °C	17 °C
Does it dissolve?	No	No	Very slightly	Slightly soluble if at all	Yes	Yes
Other observations (if any)	None	None	None	Two layers	None	None

Answers to questions

1.

Table 4.1
Properties of chlorides of Period 3

Formula of chloride	NaCl	$MgCl_2$	$AlCl_3$	$SiCl_4$	PCl_3	S_2Cl_2	Cl_2
Melting point/°C	80	714	190	–70	–112	–80	–101
Boiling point/°C	1467	1412	183	58	76	136	–34.6
Physical state at r.t.p.*	Solid	Solid	Solid	Liquid	Liquid	Liquid	Gas
$\Delta H_f^\ominus$/kJ mol^{-1}	–411	–642	–695	–640	–339	–60.2	0
$\Delta H_f^\ominus$ per mole of Cl/kJ mol^{-1}	–411	–321	–232	–160	–113	–30.1	0
Conductivity of liquid	Good	Good	Poor	Very poor	Very poor	Very poor	Very poor
Action of water	Dissolves	Dissolves	Reacts	Reacts	Reacts	Reacts	Reacts
pH of aqueous solution	7	6.5	3	2	2	2	2
Solubility in cyclohexane	Insoluble	Insoluble	Insoluble	Dissolves	Dissolves	Dissolves	Dissolves
Structure	Giant	Giant	Simple molecular†	Simple molecular	Simple molecular	Simple molecular	Simple molecular
Bonding	Ionic	Ionic	Covalent	Covalent	Covalent	Covalent	Covalent

*r.t.p. = room temperature and pressure (i.e. 25 °C and 1 atm).
†The solid has a layer structure. The vapour contains Al_2Cl_6 molecules.

2. **a** $AlCl_3(s) + 6H_2O(l) \rightarrow [Al(H_2O)_6]^{3+}(aq) + 3Cl^-(aq)$
hexaaquaaluminium(III) ion
The $[Al(H_2O)_6]^{3+}$ ions dissociate to form H_3O^+ ions giving rise to an acidic solution. There is also some HCl gas produced.

b $SiCl_4(l) + (x + 2)H_2O(l) \rightarrow SiO_2{\cdot}xH_2O(aq) + 4HCl(g)$
hydrated silicon(IV) oxide
(Some books refer to hydrated silicon(IV) oxide as silicic acid and give it the formula H_2SiO_3 or $Si(OH)_4$. These species may exist in solution but it seems likely that a range of hydrates such as $SiO_2{\cdot}2H_2O$ and $SiO_2{\cdot}H_2O$ exist which we represent as $SiO_2{\cdot}x\,H_2O$. The hydrated oxide appears as a colloidal solution or a gel.)

c $PCl_3(l) + 3H_2O(l) \rightarrow H_2PHO_3(aq) + 3HCl(aq)$
phosphonic acid
Phosphonic acid H_2PHO_3 is sometimes called phosphorous acid and given the formula H_3PO_3.

EXPERIMENT 4.13 Preparing anhydrous aluminium chloride

This experiment requires a fair degree of manipulative skill. A helping hand from you, combined with the ILPAC video programme 'Preparation of aluminium chloride' (if available), will be particularly useful to the student. Several accidents have been reported when preparing chlorine from concentrated hydrochloric acid and potassium manganate(VII) (permanganate), when concentrated sulphuric acid has been used in place of hydrochloric, with a consequential explosion. Concentrated sulphuric acid should not be in the laboratory when preparing chlorine in this way, and must certainly not be used to dry the gas.

Specimen results

Results Table 4.17

Mass of aluminium	0.25 g
Mass of empty specimen tube, m_1	10.20 g
Mass of specimen tube and product, m_2	10.60 g
Mass of product, $m = (m_2 - m_1)$	0.40 g
% yield	33%

Answers to questions

1. Any moisture present in the apparatus, such as water vapour mixed with the chlorine, would hydrolyse the aluminium chloride as it formed.
2. The product would contain aluminium oxide as an impurity. It would be formed by the reaction

$$2AlCl_3(s) + 3H_2O(g) \rightarrow Al_2O_3(s) + 6HCl(g)$$

3. Using the data from the specimen results the calculation is as follows:

$$Al(s) + 1\tfrac{1}{2}Cl_2(g) \rightarrow AlCl_3(s)$$

First, calculate the mass of $AlCl_3$ that would theoretically be formed from 0.25 g of Al

27 g of Al produces 133.5 g of $AlCl_3$

1.0 g of Al produces $\frac{133.5\text{ g}}{27}$ of $AlCl_3$

0.25 g of Al produces $\frac{133.5\text{ g}}{27} \times 0.25 = 1.2$ g of $AlCl_3$

$$\%\text{ yield} = \frac{\text{actual mass of product}}{\text{theoretical maximum mass of product}} \times 100$$

$$\therefore\ \%\text{ yield} = \frac{0.40\text{ g}}{1.2\text{ g}} \times 100 = \mathbf{33\%}$$

This experiment often gives low yields. Likely reasons are:

a Some product remains in the combustion tube and receiver bottle.
b Some product is carried away with excess chlorine.
c Some aluminium reacts with residual air to form oxide which remains in the combustion tube.
d Some product is hydrolysed by residual water vapour.

4. A desiccator is necessary to prevent hydrolysis by moist air.

EXPERIMENT 4.14 Investigating the properties of Period 3 oxides

This experiment is fairly straightforward and similar to Experiment 4.12.
When bubbling SO_2 through a measuring cylinder containing water, the students will see a significant decrease in the size of the gas bubbles as they rise to the surface of the water. The bubbles may dissolve completely before they reach the top.

Specimen results

Results Table 4.18

	Na_2O_2	MgO	Al_2O_3	SiO_2	P_4O_{10}	SO_2
Appearance	White or pale yellow solid	White solid	White solid	White solid	White solid	Colourless gas
On mixing with water						
Initial temperature	22 °C	22 °C	23 °C	23 °C	22 °C	22 °C
Final temperature	24 °C	23 °C	23 °C	23 °C	25 °C	23 °C
Does it dissolve?	Yes	Partially	No	No	Yes	Yes
pH of solution	12	9	7	7	below 4	below 4
Other observation(s) (if any)	Vigorous reaction. Colourless gas given off. Relights glowing splint	None	None	None	Vigorous reaction. Steamy fumes evolved	

Answers to questions

1.

Table 4.2

Formula of oxide	Na_2O_2*	MgO	Al_2O_3	SiO_2	P_4O_{10}*	SO_2*	Cl_2O*
Melting point/°C	460	2900	2040	1610 (quartz)	580	–75	–20
Boiling point/°C	Decomposes	3600	2980	2230 (quartz)	300	–10	2
State at s.t.p.	Solid	Solid	Solid	Solid	Solid	Gas	Gas
Action of water	Reacts	None apparent	None apparent	Insoluble	Reacts	Reacts	Reacts
pH of aq. solution	11–12	8–9	7–8	6–7	Below 3	3 or less	3 or less
Acid–base nature	Basic	Basic	Amphoteric	Acidic	Acidic	Acidic	Acidic
Conductivity of liquid	Good	Good	Good	Very poor	Very poor	Very poor	Very poor
Solubility in cyclohexane	Insoluble	Insoluble	Insoluble	Insoluble	Slightly soluble	Dissolves	Dissolves
Structure	Giant	Giant	Giant	Giant	Simple molecular	Simple molecular	Simple molecular
Bonding	Ionic	Ionic	Ionic	Covalent	Covalent	Covalent	Covalent

*These substances represent the most familiar or readily available oxides of that element.

2. **a** $Na_2O_2(s) + 2H_2O(l) \rightarrow 2NaOH(aq) + H_2O_2(aq)$

$$2H_2O_2(aq) \rightarrow 2H_2O(l) + O_2(g)$$

b Magnesium oxide reacts slightly in cold water and moderately in hot to form the hydroxide, which is very slightly soluble.

$$MgO(s) + H_2O(l) \rightarrow Mg(OH)_2(s)$$

c $P_4O_{10}(s) + 6H_2O(l) \rightarrow 4H_3PO_4(aq)$

d $SO_2(g) + H_2O(l) \rightarrow H_2SO_3(aq)$

e $Cl_2O(g) + H_2O(l) \rightarrow 2HClO(aq)$

3. The oxides change from being giant ionic structures in Groups I, II and III through a giant covalent structure for silicon(IV) oxide, to simple covalent molecular structures for phosphorus, sulphur and chlorine.
4. The acid–base nature changes from basic for the s-block oxides through amphoteric for aluminium oxide to acidic for the oxides of the remaining elements.
5. As the structure and bonding of the oxides change from giant ionic to simple molecular across the period so their acid–base nature changes from basic to acidic.

EXPERIMENT 5.1 Chemical properties of alkanes

Please see the notes on pages 4 and 5, which give details of the hazards of using organic chemicals, and describe how to dispose of them safely.

The only positive reaction, apart from combustion, in this experiment is the reaction between cyclohexane and bromine in an inert solvent in the presence of ultraviolet light. Volasil 244 is chosen as the inert solvent in which to dissolve bromine. This replaces 1,1,1-trichloroethane which is now no longer available. Details of this new solvent are given under the heading 'Use of halogenated compounds and their alternatives' on page 5.

Specimen results

Results Table 5.1

Reaction	Observations
A **Combustion**	
Appearance of flame	Orange and blue flame
Sootiness	Very little soot
B **Action of bromine** (in inert solvent)	
1. In dark	1. Liquids mix. No other change
2. In light	1. Liquids mix and decolorise. Steamy gas evolved
Identification of gas	Damp blue litmus paper turns red. White fumes with ammonia stopper. (HBr produced)
C **Action of acidified potassium manganate(VII)**	No reaction. Liquids remain separate
D **Action of concentrated sulphuric acid**	No reaction. Liquids remain separate

Answers to questions

1. **a** $\bar{E}$ (C—C) (general) = 346 kJ mol^{-1}
   $\bar{E}$ (C—H) (general) = 413 kJ mol^{-1}
   These values are higher than most of the other single bonds listed.
   b The high C—C and C—H bond energies suggest that the alkanes are potentially unreactive compounds.
   c Yes, cyclohexane fails to react with three powerful reagents, and reacts with a fourth only in the presence of light.

2. **a** Carbon dioxide and water.
   b $CH_4(g) + 2O_2(g) \rightarrow CO_2(g) + 2H_2O(l)$; $\Delta H^\circ = -890\ \text{kJ mol}^{-1}$
   $2C_2H_6(g) + 7O_2(g) \rightarrow 4CO_2(g) + 6H_2O(l)$; $\Delta H^\circ = -3120\ \text{kJ mol}^{-1}$
   or $C_2H_6(g) + 3\tfrac{1}{2}O_2(g) \rightarrow 2CO_2(g) + 3H_2O(l)$; $\Delta H^\circ = -1560\ \text{kJ mol}^{-1}$
   $C_3H_8(g) + 5O_2(g) \rightarrow 3CO_2(g) + 4H_2O(l)$; $\Delta H^\circ = -2220\ \text{kJ mol}^{-1}$
   c The highly exothermic combustion reactions of alkanes make them valuable fuels.
3. **a** In a substitution reaction, one atom (or group of atoms) in a molecule is replaced by another atom (or group).
   b

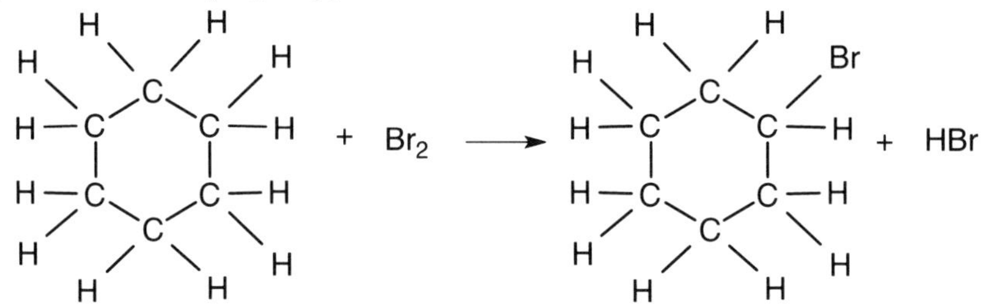

   c Ultraviolet light. ('Light' alone is **not** good enough. It must be ultraviolet light.)
   d $CH_4 + Br_2 \rightarrow CH_3Br + HBr$ bromomethane
   $CH_3Br + Br_2 \rightarrow CH_2Br_2 + HBr$ dibromomethane
   $CH_2Br_2 + Br_2 \rightarrow CHBr_3 + HBr$ tribromomethane
   $CHBr_3 + Br_2 \rightarrow CBr_4 + HBr$ tetrabromomethane

EXPERIMENT 5.2 Chemical properties of alkenes

Please see the notes on pages 4 and 5, which give details of the hazards of using organic chemicals, and describe how to dispose of them safely.

The reaction of ethene with bromine in an inert solvent is now banned as a preparation since the product, 1,2-dibromoethane, is carcinogenic; it is wise, therefore, to avoid testing ethene in this way. In the absence of any evidence about the possible dangers of 1,2-dibromocyclohexane, students should err on the side of caution and dispose of the product quickly. Aqueous bromine can be used for test purposes; the fact that a different product is formed need not be discussed unless the syllabus requires it.

Specimen results

Results Table 5.2

Reaction	Observations
A **Combustion**	
Appearance of flame	Orange and blue flame
Sootiness	Slightly sooty
B **Action of bromine** (in inert solvent)	
1. In dark	1. Liquids mix and decolorise
2. In light	2. Liquids mix and decolorise
Identification of gas	No gas is given off
C **Action of acidified potassium manganate(VII)**	Rapid decolorisation. Liquids remain separate
D **Action of concentrated sulphuric acid**	Liquids mix and react vigorously. Colour darkens to charred black mass

Answers to questions

1. **a** $-\Delta H^\circ \simeq \bar{E}(C{-}C) + 6\bar{E}(C{-}H)$
   $\therefore \bar{E}(C{-}C) = -\Delta H^\circ - 6\bar{E}(C{-}H)$
   $= 2820\ \text{kJ mol}^{-1} - (6 \times 412)\ \text{kJ mol}^{-1} = \mathbf{348\ kJ\ mol^{-1}}$

   b $-\Delta H^\circ \simeq \bar{E}(C{=}C) + 4\bar{E}(C{-}H)$
   $\therefore \bar{E}(C{=}C) = -\Delta H^\circ - 4\bar{E}(C{-}H)$
   $= 2260\ \text{kJ mol}^{-1} - (4 \times 412)\ \text{kJ mol}^{-1} = \mathbf{612\ kJ\ mol^{-1}}$

   c $\bar{E}(C{=}C) = \bar{E}(\sigma) + \bar{E}(\pi)$
   $\therefore \bar{E}(\pi) = \bar{E}(C{=}C) - \bar{E}(\sigma)$
   $= 612\ \text{kJ mol}^{-1} - 348\ \text{kJ mol}^{-1} = \mathbf{264\ kJ\ mol^{-1}}$
   This is about $^3/_4$ of $\bar{E}(\sigma)$.

   d It requires less energy to break the π-bond in ethene than the σ-bond in ethane. This suggests that ethene might be more reactive than ethane.

   e Yes. A comparison of the results for Experiments 5.1 and 5.2 shows more positive reactions for cyclohexene than cyclohexane, indicating the greater reactivity of alkenes in general.
2. The presence of soot in the flame is an indication of the higher carbon to hydrogen ratio in alkenes than in alkanes. Each C atom requires two O atoms for complete combustion whereas two H atoms require only one O atom.
3. Unlike alkanes, alkenes decolorise bromine in the absence of light and acidified potassium manganate(VII) solution.

EXPERIMENT 5.3 Hydrolysis of organic halogen compounds

Please see the notes on pages 4 and 5, which give details of the hazards of using organic chemicals, and describe how to dispose of them safely.

The order of ease of hydrolysis in this experiment is 1-iodobutane > 1-bromobutane > 1-chlorobutane > chlorobenzene.

Ethanol is added in this experiment as a common solvent and to prevent the halogeno-compounds forming an emulsion with the water present, since this can be confused with the formation of a precipitate of silver halide.

The experiment can easily be adapted to show the different rates of hydrolysis of primary, secondary and tertiary halogenoalkanes.

Specimen results

Results Table 5.3

Reaction	Time for precipitate to appear	Observations
A 1-Chlorobutane	8 minutes	Faint still after 30 minutes
B 1-Bromobutane	2 minutes	Denser after 5 minutes. Settling beginning after 30 minutes
C 1-Iodobutane	30 seconds	Heavy yellow after 1 minute. Settled after 30 minutes
D Chlorobenzene		No precipitate even after 1 hour

Answers to questions

1. The trend in bond energy values suggests that we can list the halogeno-compounds with the most reactive first in the order:

   $$C_4H_9I > C_4H_9Br > C_4H_9Cl > C_6H_5Cl$$

2. 1-Iodobutane, 1-bromobutane, 1-chlorobutane, chlorobenzene.

3. Yes, the list predicting potential reactivity coincides with the order of speed of hydrolysis.

4. $$CH_3CH_2CH_2CH_2X + H_2O \rightarrow \underset{\text{butan-1-ol}}{CH_3CH_2CH_2CH_2OH} + H^+(aq) + X^-(aq) \quad (X = Cl, Br \text{ or } I)$$

EXPERIMENT 5.4 Preparation of a halogenoalkane

Please see the notes on pages 4 and 5, which give details of the hazards of using organic chemicals, and describe how to dispose of them safely.

We have chosen the preparation of 2-chloro-2-methylpropane because it is simpler and more readily controlled than the widely used preparation of 1-bromobutane.

We suggest you spend some time instructing your students on the assembly, care and cleaning of ground-glass-joint apparatus, especially if this is the first time they use it. They may also need help with the techniques of simple distillation and the use of a separating funnel. The ILPAC video programme 'Organic Techniques I', if available should prove very useful in all these areas. You may think it worth purchasing plastic clips (e.g. Griffin, QJM 200-W) to hold joints firmly together.

Students should not be discouraged by low yields at this stage. You should discuss with them possible reasons for low yields.

Specimen results

Results Table 5.4

Mass of measuring cylinder + 2-methylpropan-2-ol	50.8 g
Mass of measuring cylinder after emptying	43.7 g
Mass of 2-methylpropan-2-ol	7.1 g
Mass of collecting flask	32.2 g
Mass of collecting flask + 2-chloro-2-methylpropane	37.7 g
Mass of 2-chloro-2-methylpropane	5.5 g

Answers to questions

1. $$\text{Amount of } (CH_3)_3COH \text{ used} = \frac{7.1\text{ g}}{74\text{ g mol}^{-1}} = 0.096\text{ mol}$$

The equation shows 1 mol of $(CH_3)_3CCl$ formed from 1 mol of $(CH_3)_3COH$
$\therefore$ the maximum possible amount of $(CH_3)_3CCl$ = 0.096 mol
and the maximum mass = 0.096 mol × 92.5 g mol^{-1} = **8.9 g**

2. $$\%\text{ yield} = \frac{\text{actual mass of product}}{\text{maximum mass of product}} \times 100 = \frac{5.5\text{ g}}{8.9\text{ g}} \times 100 = \mathbf{62\%}$$

3. A stronger alkali would tend to hydrolyse the halogeno-compound back to the alcohol.

EXPERIMENT 6.1 The effect of concentration changes on equilibria

This is a very simple experiment and should take only a few minutes. We suggest that students should not get involved at this stage in discussing the various possible formulae of the complex ion(s) formed by iron(III) in the presence of excess chloride ions. For this reason, we have written an equation in the simplest possible form.

Specimen results
Results Table 6.1

Change	Observation	Cause	Inference
$[Fe^{3+}]$ increased	Darker colour	Concentration of $FeSCN^{2+}$ increases	Equilibrium shifts to the right
$[SCN^-]$ increased	Darker colour	Concentration of $FeSCN^{2+}$ increases	Equilibrium shifts to the right
$[Fe^{3+}]$ decreased	Lighter colour	Concentration of $FeSCN^{2+}$ decreases	Equilibrium shifts to the left

1. The position of equilbrium would shift to the left.
2. **a** The system counteracts the imposed change (an increase in $[Fe^{3+}]$) by converting some Fe^{3+} and some SCN^- to $FeSCN^{2+}$.
   b The system counteracts the imposed change (an increase in $[SCN^-]$) by converting some SCN^- and some Fe^{3+} to $FeSCN^{2+}$.
   c The system counteracts the imposed change (a decrease in $[Fe^{3+}]$) by converting $FeSCN^{2+}$ to Fe^{3+} and SCN^-.

EXPERIMENT 6.2 Determining an equilibrium constant

This experiment can give excellent results if the procedure is followed carefully. However, students may need help in understanding the procedure if it is not to become a mechanical exercise. Weaker students especially will need guidance through the rather lengthy calculations. If you have a suitable water-bath, you may wish to introduce temperature control, but we do not specify this because the equilibrium constant varies very little over the normal range of room temperatures. Carefully standardised sodium hydroxide solution is needed for this experiment.

It may be worthwhile showing that equilibrium can also be reached from the 'other end' by weighing out ethanoic acid and ethanol instead of ester and water. The combined volume of acid and alcohol should again be 5 cm^3, using, say, 3 + 2, 2.5 + 2.5 and 2 + 3 cm^3 of the liquids.

Students may find the calculations difficult and time-consuming. However, so long as they understand the calculation **method,** the detailed working can be performed by a computer spreadsheet which saves valuable time.

Specimen results
Results Table 6.2

Tube number	1a	1b	2	3	4
Mass of empty tube/g	12.01	12.24	12.57	12.18	12.73
Volume of HCl(aq) added/cm^3	5.0	5.0	5.0	5.0	5.0
Mass of tube after addition/g	17.03	17.34	17.77	17.35	17.92
Volume of ethyl ethanoate added/cm^3	–	–	5.0	4.0	2.0
Mass of tube after addition/g			22.28	20.99	19.78
Volume of water added/cm^3	–	–	–	1.0	3.0
Mass of tube after addition/g				21.98	22.78
Mass of ethyl ethanoate added/g			4.51	3.64	1.86
Mass of HCl(aq) added/g			5.20	5.17	5.19
Mass of water added/g				0.99	3.00

Results Table 6.3

Solution in flask	Equilibrium mixture				
Solution in burette	Sodium hydroxide 0.974 mol dm^{-3}				
Indicator	Phenolphthalein				
Tube number	1a	1b	2	3	4
Final burette reading	9.95	20.05	41.30	39.20	27.25
Initial burette reading	0.00	9.95	0.00	0.00	0.00
Titre/cm^3	9.95	10.10	41.30	39.20	27.25

Results Table 6.4

Tube number	**2**	**3**	**4**
1. Amount of HCl/mol	0.0100	0.0100	0.0100
2. Total amount of acid at eqm/mol	0.0413	0.0392	0.0273
3. Eqm amount of ethanoic acid/mol	0.0313	0.0292	0.0173
4. Eqm amount of ethanol/mol	0.0313	0.0292	0.0173
5. Initial amount of ethyl ethanoate/mol	0.0511	0.0413	0.0211
6. Eqm amount of ethyl ethanoate/mol	0.0198	0.0121	0.0038
7. Mass of pure HCl/g	0.365	0.365	0.365
8. Mass of water in HCl(aq)/g	4.84	4.81	4.83
9. Initial amount of water/mol	0.269	0.322	0.435
10. Eqm amount of water/mol	0.238	0.293	0.418
11. Eqm constant, K_c	0.204	0.240	0.19

Answers to questions

1. According to Le Chatelier's principle, increasing the concentration of a reactant and decreasing the concentration of a product should both result in the shifting of the equilibrium position to the right. Therefore both the addition of water and titration with alkali should result in more ethanoic acid and ethanol being formed.
2. Dilution of the equilibrium mixture with water slows down both the forward and backward reactions (sometimes this is referred to as 'freezing the equilibrium'). Therefore the equilibrium position changes so slowly that there is time to complete the titration before the equilibrium is disturbed.

EXPERIMENT 6.3 Determining a solubility product

Calcium hydroxide is too soluble for the solubility product, in terms of concentrations rather than activities, to be truly constant, and this is why some data books do not list a value. However, a simple titration method cannot be readily applied to less soluble salts, and this experiment does give quite good results. This experiment is suitable for a practical assessment exercise, so a suggested marking scheme and specimen results follow.

Suggested mark scheme

Total marks for manipulation = 12, as follows:

Measurement of approx. 2 g of $Ca(OH)_2$ (1)
Correct use of measuring cylinder (1)
Shaking of stoppered bottles (1)
Correct filtering of suspensions (1)

Rejection of first 5 cm^3 of filtrate	(1)
Correct filling of burette	(1)
Correct use of pipette filler	(1)
Correct use of pipette	(1)
Correct titration technique	(1)
Correct reading of burette	(1)
Carries out one trial and at least two accurate titrations	(1)
Recording of temperature	(1)

Total mark	12–11	10–9	8–6	5–4	3–2	1–0
Scaled mark	5	4	3	2	1	0

Total marks for interpretation = 10, as follows:

Calculation of $[OH^-]$ in saturated $Ca(OH)_2(aq)$	(3)
Calculation of $[Ca^{2+}]$ in saturated $Ca(OH)_2(aq)$	(1)
Calculation of solubility of $Ca(OH)_2$ from experimental result	(1)
Comparison with value from data book	(1)
Calculation of K_{sp} from experimental result (+ correct units)	(2)
Calculation of K_{sp} from data book (+ correct units)	(2)

Divide total by two to obtain scaled mark.

Specimen results

Results Table 6.5

Solution in flask	Calcium hydroxide			? mol dm^{-3}		25.0 cm^3
Solution in burette	Hydrochloric acid			0.100 mol dm^{-3}		
Indicator	Phenolphthalein					
		Trial	1	2	3	4
Burette readings	Final		11.60	23.00	34.45	46.00
	Initial		0.00	11.60	23.00	34.50
Volume used/cm^3			11.60	11.40	11.45	11.50
Mean titre/cm^3	11.45		Temperature = 15 °C			

Calculation

1. Let A refer to $H^+(aq)$ and B refer to $OH^-(aq)$

$$\frac{c_A V_A}{c_B V_B} = \frac{1}{1}$$

$$\frac{0.100 \text{ mol dm}^{-3} \times 11.45 \text{ cm}^3}{c_B \times 25.0 \text{ cm}^3} = 1$$

$$\therefore\ c_B = \frac{0.100 \text{ mol dm}^{-3} \times 11.45 \text{ cm}^3}{25.0 \text{ cm}^3} = \mathbf{0.0458\ mol\ dm^{-3}}$$

2. $[Ca^{2+}(aq)] = \frac{1}{2}[OH^-(aq)]$ = **0.0229 mol dm^{-3}**
3. Solubility of $Ca(OH)_2$ = $[Ca^{2+}(aq)]$ = 0.0229 mol dm^{-3} (at 15 °C)
   Two different data books list solubilities in different ways:
   a 0.156 g of $Ca(OH)_2$ per 100 g of H_2O at 20 °C
   or 1.56 g dm^3 or 1.56/74.0 mol dm^{-3} = **0.0211 mol dm^{-3}**
   b **1.53 × 10^{-3} mol per 100 g of H_2O** at 25 °C.
   Note that solubilities are usually quoted with reference to the volume of **water** rather than the volume of **solution**. This makes very little difference for dilute solutions such as saturated calcium hydroxide. Also, note that calcium hydroxide unlike most salts, becomes **less** soluble at higher temperatures.

4. **a** $K_s = [Ca^{2+}(aq)][OH^-(aq)]^2$
$= 0.0229 \text{ mol dm}^{-3} \times (0.0458 \text{ mol dm}^{-3})^2 = \mathbf{4.80 \times 10^{-5} \text{ mol}^3 \text{ dm}^{-9}}$
b $K_s = 0.0211 \text{ mol dm}^{-3} \times (0.0422 \text{ mol dm}^{-3})^2 = \mathbf{3.76 \times 10^{-5} \text{ mol}^3 \text{ dm}^{-9}}$ or
$K_s = 0.0153 \text{ mol dm}^{-3} \times (0.0306 \text{ mol dm}^{-3})^2 = \mathbf{1.43 \times 10^{-5} \text{ mol}^3 \text{ dm}^{-9}}$
(Another data book lists the solubility product at 25 °C as 5.5×10^{-6} $\text{mol}^3 \text{ dm}^{-9}$. Clearly there is a considerable degree of uncertainty.)

EXPERIMENT 6.4 Illustrating the common ion effect

This is a simple experiment and should take only a few minutes. Note that a soluble salt such as sodium chloride can be used to illustrate the common ion effect even though the solubility product principle cannot be applied (in terms of concentrations) because of the interaction between ions at high concentrations.

Answers to questions

1. A colourless precipitate appeared in each tube.
2. According to Le Chatelier's principle, the system counteracts an increase in the concentration of either of the ions by shifting the equilibrium position to the left (i.e. producing solid NaCl).

$$NaCl(s) \rightleftharpoons Na^+(aq) + Cl^-(aq)$$

3. Each ion has concentration = 5.42 mol dm^{-3}.
4. Some of the sodium ions in solution interacted with the excess chloride ions to form a precipitate of sodium chloride.
5. Some of the chloride ions in solution interacted with the excess sodium ions to form a precipitate of sodium chloride.
6. At high concentrations, ions of opposite charge interact with each other, tending to 'pair up'. This makes the effective concentration of 'free' ions very much smaller than the concentrations obtained from the amount dissolved. The equilibrium law can only be applied using concentrations if these are the same as the effective concentrations of 'free' ions, which is the case in very dilute solutions only. (Effective concentrations are known as 'activities' and, if values can be calculated, they can be used in the equilibrium law, but this is beyond A-level.)

EXPERIMENT 6.5 Distribution equilibrium

Most chlorinated hydrocarbons are harmful or toxic and many of them are depleters of the ozone layer. In the first edition of this book the solvent specified for this experiment was 1,1,1-trichloroethane but this became unavailable to schools from 1 January 1995. We have therefore replaced it by Volasil 244, obtainable from BDH or from Merck. This cyclic silicone is flammable but it is non-toxic.

Although we have recommended that both hydrochloric acid solutions should be standardised, you may feel that it is sufficient to make up approximately 0.5 M HCl and then dilute this carefully so that the ratio of concentrations is precisely 50:1. The end-point may not be very sharp using 0.010 M HCl; you may wish to warn your students of this.

Specimen results

Results Table 6.6

Solution in flask		NH_3 (aq)		? mol dm^{-3}		10 cm^3
Solution in burette		HCl (aq)		0.50 mol dm^{-3}		
Indicator		Methyl red				
		Trial	1	2	3	4
Burette readings	Final		17.15	34.10	16.80	
	Initial		0.00	17.15	0.00	
Volume used/cm^3			17.15	16.95	16.80	
Mean titre/cm^3	16.97					

Results Table 6.7

Solution in flask		NH_3 (Volasil)		? mol dm^{-3}		10 cm^3
Solution in burette		HCl (aq)		0.010 mol dm^{-3}		
Indicator		Methyl red				
		Trial	1	2	3	4
Burette readings	Final		5.80	12.00	17.90	
	Initial		0.00	5.80	12.00	
Volume used/cm^3			5.80	6.20	5.90	
Mean titre/cm^3	5.97					

Calculation

1. Substituting in the expression
   $c_A V_A / c_B V_B = a/b$ where A refers to HCl and B to NH_3

   gives
   $$\frac{0.50 \text{ mol dm}^{-3} \times 16.97 \text{ cm}^3}{c_B \times 10.0 \text{ cm}^3} = \frac{1}{1}$$

   $\therefore c_B = 0.50 \text{ mol dm}^{-3} \times 16.97/10.0 = \mathbf{0.85 \text{ mol dm}^{-3}}$
2. By the same method as in 1.
   $c_B = 0.010 \text{ mol dm}^{-3} \times 5.97/10.0 = \mathbf{6.0 \times 10^{-3} \text{ mol dm}^{-3}}$

3. $K_D = [NH_3(aq)]/[NH_3(Vol)]$
   $= 0.85 \text{ mol dm}^{-3}/6.0 \times 10^{-3} \text{ mol dm}^{-3} = \mathbf{140}$

EXPERIMENT 7.1 The pH of a weak acid at various concentrations

See notes for Experiment 7.2 below.

Specimen results

Results Table 7.1

Concentration of acid/mol dm^{-3}	**Observed pH of solutions of ethanoic acid**	**Calculated pH of solutions of hydrochloric acid**
0.000 10	4.2	4.0
0.0010	3.5	3.0
0.010	3.0	2.0
0.10	2.7	1.0

Answers to questions

1. **a** In each case, the hydrochloric acid solution has a higher concentration of hydrogen ions than the ethanoic acid solution of the same concentration.
   b This shows that the ethanoic acid solutions are incompletely dissociated.
2. **a** As the ethanoic acid is diluted, its pH becomes closer and closer to the pH of the strong acid at the same concentration. This shows that dilution increases the extent of dissociation of ethanoic acid. The same is true of other weak acids.
   The trend is continued at greater dilution, but precise measurement becomes much more difficult due to contamination of the sample with dissolved CO_2.
   However, calculations show that even at a concentration of 10^{-6} mol dm^{-3}, the ethanoic acid would not be quite fully dissociated; it would have a pH of 6.02 as compared with pH of 6.00 for a strong acid.
   b The full equation for the dissociation of ethanoic acid in water is:

$$CH_3CO_2H(aq) + H_2O(l) \rightleftharpoons CH_3CO_2^-(aq) + H_3O^+(aq)$$

The addition of water decreases the concentrations of ethanoic acid, ethanoate ions and hydrogen ions. Because two of these are on the right of the equation and only one on the left, a shift to the right goes further towards restoring the original concentrations than a shift to the left. The decrease in concentration reduces the rate of reaction R $\rightarrow$ L to a greater extent than L $\rightarrow$ R. Thus, the acid becomes increasingly dissociated on dilution.

Note: the alternative simpler argument, that the added water is partially used up in a shift to the right, is invalid because the concentration of **water** remains effectively constant.

EXPERIMENT 7.2 The pH of different acids at the same concentration

These experiments take only a few minutes, and one pH meter can easily serve a class. We suggest that, at this stage, you teach your students to use a pH meter as a tool without considering how it works; a brief account can be given, if required, during the study of redox reactions.

We think it best not to measure the pH of hydrochloric acid for comparison purposes; meters actually measure −log (activity) and, consequently, 0.1 M HCl appears to have a pH greater than 1.

If a pH meter is not available, you may wish to try using narrow-range pH papers, but they are not very satisfactory over this range.

Specimen results

Results Table 7.2

Acid	**Concentration /mol dm^{-3}**	**pH**
Boric acid	0.010	5.6
Benzoic acid	0.010	2.8
Ethanoic acid	0.010	3.0
Dihydrogenphosphate(V) ion	0.010	4.6

Answer to question

Order of acid strength, strongest acid first:

benzoic acid > ethanoic acid > dihydrogenphosphate(V) ion > boric acid

EXPERIMENT 7.3 The action of a buffer solution

Students may not observe pH changes equal to those we have quoted in our specimen results, but the contrast between the buffer and water should be quite clear. If you use commercial buffer tablets, we suggest you check that the solution is concentrated enough to absorb the amounts of acid and alkali we have specified – it may be necessary to dilute the NaOH and HCl solutions. The buffer recipe we have quoted contains more HPO_4^{2-} than $H_2PO_4^-$; consequently, it absorbs acid more effectively than alkali.

Students may have some difficulty in measuring the pH of pure water quickly enough before absorption of carbon dioxide decreases the value.

Specimen results

Results Table 7.3

Volume added	**pH on addition of 0.1 M NaOH to**		**pH on addition of 0.1 M HCl to**	
	Buffer	**Pure water**	**Buffer**	**Pure water**
0	7.0	6.9	7.0	6.8
1 drop	7.0	9.8	7.0	4.1
1.0 cm^3	7.1	10.5	6.9	2.7
5.0 cm^3	10.0	11.6	6.0	2.1
pH of pure water with minimum air exposure		6.9		
pH of pure water after 10 minutes air exposure		5.4		

Answers to questions

1. Students will probably record a change of about 4 pH units. Theoretically, the change should be from pH 7.0 to pH 2.4 so, if everything is right in their experiments, students might answer 5 pH units.
2. Assuming that the pH of pure water after the addition of acid was 3.0, to the nearest unit:

$$[H^+(aq)] = 1.0 \times 10^{-3}\ \text{mol dm}^{-3}$$
$$[H^+(aq)] \text{ in pure water (pH 7.0)} = 1.0 \times 10^{-7}\ \text{mol dm}^{-3}$$
$$\therefore \frac{[H^+(aq)] \text{ after addition of acid}}{[H^+(aq)] \text{ in pure water}} = \frac{1.0 \times 10^{-3}\ \text{mol dm}^{-3}}{1.0 \times 10^{-7}\ \text{mol dm}^{-3}} = \mathbf{10^4}$$

   In other words, a decrease of 4 pH units means there are 10^4 (or 10 000) times as many hydrogen ions in the same volume.
3. Almost all the hydrogen ions must have reacted with the buffer solution.
4. Carbon dioxide is absorbed from the air and reacts with water to form hydrogen ions.

$$CO_2(aq) + H_2O(l) \rightleftharpoons H^+(aq) + HCO_3^-(aq)$$

   The distilled water normally used in the laboratory is pure enough for most purposes, but nevertheless it rapidly becomes somewhat acidic.

EXPERIMENT 7.4 Determining the pH range of some acid–base indicators

We suggest that results are shared amongst class members so that individual students do not spend too much time on this experiment. Make sure that students understand the procedure, particularly the need to have the same **concentration** of indicator in the solutions being compared.

The universal buffer solution supplied by BDH Laboratory Supplies contains phenylethanoic acid, which has a most unpleasant smell! Students should be warned to keep solutions stoppered or covered and to neutralise waste with an alkali. The alternative contains ethanoic acid instead and, although its initial pH is slightly higher (3.25), the approximate relationship between pH and volume of added alkali still holds.

Specimen results

Results Table 7.4

Indicator used	Initial colour	Colour change starts		Colour change ends		Final colour
		Volume	pH	Volume	pH	
Litmus	Red	8 cm^3	4.9	24 cm^3	8.5	Blue
Methyl orange	Red	1 cm^3	3.0	11 cm^3	5.6	Yellow
Phenolphthalein	Colourless	22 cm^3	8.0	30 cm^3	9.9	Red
Bromophenol blue	Yellow	1 cm^3	3.0	9 cm^3	5.1	Blue
Methyl red	Red	8 cm^3	4.9	15 cm^3	6.4	Yellow

Table 7.1

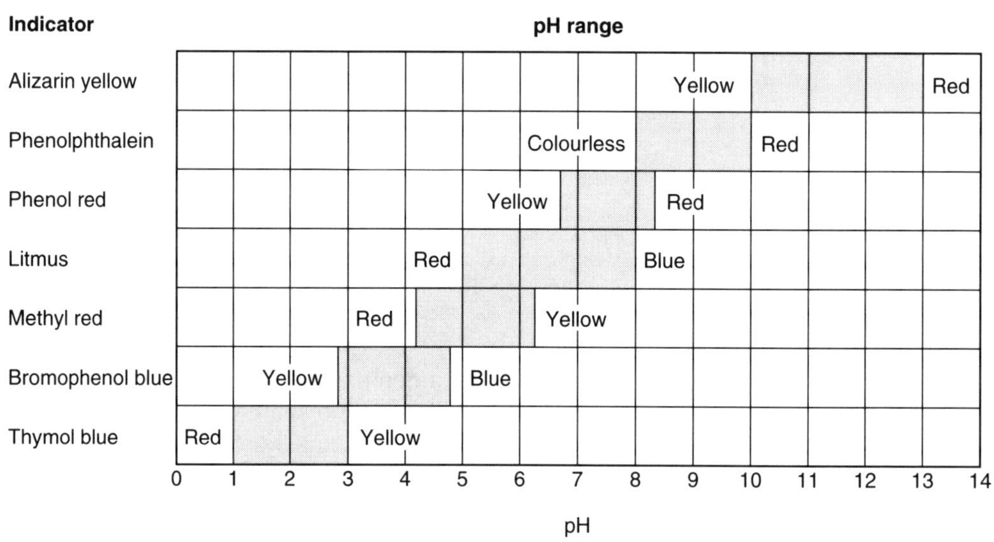

Answer to question

The pH was measured in a buffer solution to ensure that the pH did not change too rapidly on the addition of alkali. It would be almost impossible to change the pH of the solution in a controlled way without using a buffer.

EXPERIMENT 7.5 Determining the ionisation constant for an indicator

It is important that all test-tubes should have the same bore. There will inevitably be some slight variations but as long as the tubes seem to match well, experimental results should be reasonably good.

A ratio of 2 : 1 for $[HIn(aq)] : [In^-(aq)]$ gives $pK_{In} = 4.0$. Students should obtain a good result because ratios as different as 3 : 1 and 1 : 1 give pK_{In} values of 4.2 and 3.7 respectively.

Specimen results and calculation

Best colour match with tubes corresponding to $\frac{[HIn(aq)]}{[In^-(aq)]} = \frac{7}{3}$

$$pH = pK_{In} - \log \frac{[HIn(aq)]}{[In^-(aq)]}$$

$$\therefore pK_{In} = pH + \log \frac{[HIn(aq)]}{[In^-(aq)]} = 3.7 + \log \frac{7}{3} = 3.7 + 0.37 = 4.1$$

$$K_{In} = \text{antilog}(-pK_{In}) = 10^{-4.1} = \mathbf{7.9 \times 10^{-5}\ mol\ dm^{-3}}$$

Answers to questions

1. The total concentrations of indicator, $[HIn(aq)]$ and $[In^-(aq)]$, had to be the same in the solutions being compared, otherwise a change in density might be confused with a change in colour. Each **pair** of tubes had a total of 10 drops of indicator.
2. A buffer of pH 8 could not be used since pH 8 is outside the pH range of the indicator. Colour comparisons can only be made when the ratio $[HIn(aq)]/[In^-(aq)]$ is between 10 and 0.1 (approximately).

EXPERIMENT 7.6 Obtaining the pH curve for an acid–alkali titration

You may choose to use this experiment for practical assessment. The specimen results below are those that should be expected if precisely 0.100 M solutions are used (which need not necessarily be the case). If the solutions are not precisely 0.100 M then you may wish to check the end-point by carrying out the first part of the procedure yourself but this will not be necessary if you are not going to assess the students' accuracy (manipulation). A suggested marking scheme for observing and recording and for interpretation appears below.

Suggested mark scheme

Total marks for observing and recording = 8, as follows:

Records titration volume V using phenolphthalein	(1)
Records results of pH titration clearly in a suitable table	(1)
Records pH readings to nearest 0.1	(1)
Records burette readings to nearest 0.1 cm^3	(1)
Records colour of solution appropriately	(1)
Records sufficient number of readings near the equivalence point	(1)
Continues recording readings up to 5 cm^3 beyond equivalence point	(1)
Observes need for stirring and waiting for pH to steady before taking a reading	(1)

Total mark	8–7	6–5	4	3–2	1	0
Scaled mark	5	4	3	2	1	0

Total marks for interpretation = 13, as follows:

Plots pH (vertical axis) against volume of NaOH (horizontal axis)	(1)
Labels axes correctly	(1)
Chooses suitable scale for each axis	(1)
Plots points correctly	(1)
Draws best curve	(1)

Determines colour change range of indicator	(1)
Comments appropriately on the suitability of the indicator	(1)
Realises half-way up the steepest portion of the curve represents the equivalence point	(1)
Records volume at the equivalence point	(1)
Recognises method/equations for calculating molarity of the acid	(1)
Calculates the molarity of the ethanoic acid	(1)
From graph obtains, and records, the pH value half-way to the equivalence point	(1)
Realises the special property of this solution (buffer)	(1)

Total mark	13–12	11–10	9–7	6–4	3–2	1–0
Scaled mark	5	4	3	2	1	0

Specimen results

Results Table 7.5

Burette reading/cm^3	pH	Colour of solution
0.0	3.0	Yellow
5.0	4.2	"
10.0	4.6	"
15.0	4.9	"
20.0	5.3	"
21.0	5.4	"
22.0	5.5	"
23.0	5.7	"
24.0	6.2	"
24.2	6.4	"
24.4	6.6	"
24.6	6.9	"
24.8	7.2	Yellow–orange
24.9	7.6	Orange
24.95	8.1	Orange–red
25.0	9.0	Red
25.05	9.5	"
25.1	10.0	"
25.2	10.6	"
25.4	10.9	"
25.6	11.4	"
25.8	11.3	"
26.0	11.4	"
30.0	11.9	"

EXPERIMENT 7.7 Preparation of buffers

You may choose to use this experiment to test planning skills as part of a practical assessment. A suggested marking scheme is given below. Students may need help in calculating the volumes of solutions to use and in understanding why the pH of the buffer does change a little on dilution when the equation they have used suggests that the pH should remain constant. The solutions we have specified give concentrated buffers with a very high buffering capacity; you may choose to use 0.1 M solutions for reasons of economy.

Suggested mark scheme

Total marks for planning = 9, as follows:

Chooses suitable solutions for preparation of buffer pH 5.2	(1)
Chooses suitable solutions for preparation of buffer pH 8.8	(1)
Calculates volumes needed for preparation of buffer pH 5.2	(1)
Calculates volumes needed for preparation of buffer pH 8.8	(1)
Chooses suitable method for measuring volumes	(1)
Chooses suitable method for measuring pH	(1)
Calculates suitable volumes for testing the dilution effect	(1)
Writes clear instructions	(1)
Prepares suitable table for recording results	(1)

Total mark	9	8–7	6–5	4–3	2–1	0
Scaled mark	5	4	3	2	1	0

Answers to questions

1. From the substances available, the acid with a pK_a value nearest to the required pH of 5.2 is ethanoic acid, pK_a = 4.8.

$$\mathrm{pH} = \mathrm{p}K_a - \log \frac{[\mathrm{CH_3CO_2H(aq)}]}{[\mathrm{CH_3CO_2^-(aq)}]}$$

$$\therefore \frac{[\mathrm{CH_3CO_2H(aq)}]}{[\mathrm{CH_3CO_2^-(aq)}]} = \text{antilog}\ (\mathrm{p}K_a - \mathrm{pH}) = \text{antilog}\ (4.8 - 5.2)$$

$$= \text{antilog}\ (-0.4) = 0.40$$

If x cm^3 of 1.0 M CH_3CO_2H is used to make 100 cm^3 of buffer, $(100 - x)$ cm^3 of 1.0 M CH_3CO_2Na is required.

$$\frac{x}{100 - x} = 0.4 \text{ or } x = 40 - 0.4\,x \qquad \therefore x = \frac{40}{1.4} = 29$$

∴ **use 29 cm^3 of 1.0 M CH_3CO_2H and 71 cm^3 of 1.0 M CH_3CO_2Na**

2. From the substances available, the acid with a pK_a value nearest to the required pH of 8.8 is the ammonium ion, pK_a = 9.3.

$$\frac{[\mathrm{NH_4^+(aq)}]}{[\mathrm{NH_3(aq)}]} = \text{antilog}\ (\mathrm{p}K_a - \mathrm{pH}) = \text{antilog}\ (9.3 - 8.8)$$

$$= \text{antilog}\ (0.5) = 3.2$$

If x cm^3 of 1.0 M NH_4Cl is used to make 100 cm^3 of buffer, $(100 - x)$ cm^3 of 1.0 M NH_3 is required.

$$\frac{x}{100 - x} = 3.2 \text{ or } x = 320 - 3.2\,x \qquad \therefore x = \frac{320}{4.2} = 76$$

∴ **use 76 cm^3 of 1.0 M NH_4Cl and 24 cm^3 of 1.0 M NH_3**

3. Measured values should be within 0.2 unit of the required values. Differences may be due to inaccurate pH meters but, in any case, the solutions used are too concentrated for the equilibrium law to be applied precisely using concentrations. pH meters actually measure −log (activity of hydrogen ions), where 'activity' is the technical term for 'effective concentration'. At high concentrations, ions interact with each other so that their effective concentration is reduced – as a result, pH meters tend to record higher values than you might expect for concentrated solutions.
4. The equation used in the calculation suggests that the pH depends only on the **ratio** of concentrations of acid and conjugate base, so that dilution should have no effect. In practice, the pH of these buffers falls a little with dilution – between 0.1 and 0.2 unit for a 10 : 1 dilution. Again, this is due to the fact that measured pH depends on activity, which changes with concentration, and also to the breakdown of the assumptions made in the calculation, e.g.
   $[CH_3COOH(aq)] \simeq$ molarity in dilute solutions
5. The capacity of the buffers falls dramatically with dilution. The concentrated buffers will absorb 10 cm^3 of 0.2 M HCl or 0.2 M NaOH with a pH change of only 0.1 or 0.2 unit. A similar change for buffers diluted by 10 : 1 is brought about by only 1 cm^3 of acid or alkali; while for buffers diluted by 100 : 1 a single drop of acid or alkali is enough.
6. Ammonia is so volatile that the composition of the buffer will not remain constant.

EXPERIMENT 7.8 Determining the dissociation constant of a weak acid

The procedure is very similar to that for Experiment 7.5, and need not be attempted if time is short. Again, students should obtain good results.

Specimen results and calculations

1. Best colour match with tubes corresponding to $\dfrac{[HIn(aq)]}{[In^-(aq)]} = \dfrac{4}{6}$

$$\therefore \text{pH} = \text{p}K_{In} - \log \frac{[HIn(aq)]}{[In^-(aq)]} = 4.0 - \log \frac{4}{6} = \mathbf{4.2}$$

2.
$$\text{pH} = \text{p}K_a - \log \frac{[C_6H_5CO_2H(aq)]}{[C_6H_5CO_2^-(aq)]}$$

$$4.2 = \text{p}K_a - 0 \quad (\text{since } [C_6H_5CO_2H(aq)] = [C_6H_5CO_2^-(aq)])$$

$$\therefore \text{p}K_a = 4.2 \quad \text{and} \quad K_a = 10^{-4.2} = \mathbf{6.3 \times 10^{-5}\ mol\ dm^{-3}}$$

EXPERIMENT 7.9 Some simple redox reactions

You may feel your students are already familiar with this simple experiment and may omit it, but we suggest that they do answer the questions. Remind your students not to wash away solutions containing silver salts – this applies to Experiment 7.10 as well.

Specimen results
Results Table 7.6

Solution	Observations		
	Zn	**Cu**	**Ag**
$Zn^{2+}(aq)$		No change	No change
$Cu^{2+}(aq)$	The zinc became covered with a brown coating of copper. The blue solution became almost colourless		No change
$Ag^{+}(aq)$	The zinc became covered with a silver-grey coat of metallic silver	The copper became covered with a silver-grey coat of metallic silver	

Answers to questions

1. **a** $Zn(s) + Cu^{2+}(aq) \rightarrow Zn^{2+}(aq) + Cu(s)$
   b $Zn(s) + 2Ag^{+}(aq) \rightarrow Zn^{2+}(aq) + 2Ag(s)$
   c $Cu(s) + 2Ag^{+}(aq) \rightarrow Cu^{2+}(aq) + 2Ag(s)$
2. **a** Zinc is oxidised to zinc ions ($0 \rightarrow +2$)
   Copper ions are reduced to copper ($+2 \rightarrow 0$)
   b Zinc is oxidised to zinc ions ($0 \rightarrow +2$)
   Silver ions are reduced to silver ($+1 \rightarrow 0$)
   c Copper is oxidised to copper ions ($0 \rightarrow +2$)
   Silver ions are reduced to silver ($+1 \rightarrow 0$)
3. Copper. In 1**a** copper is reduced whilst in 1**c** it is oxidised.

EXPERIMENT 7.10 Measuring the potential difference generated by electrochemical cells

This is a simple experiment but it is most important for what follows that your students understand how the polarity of the electrodes enables them to determine what reaction is occurring.

Specimen results
Results Table 7.7

Cell	Positive electrode	Negative electrode	Potential difference/V
1	Copper	Zinc	1.06
2	Silver	Copper	0.40
3	Silver	Zinc	1.48

Students' readings may differ from these by up to 10% because the conditions of measurement are not standardised.

Answers to questions

1. **a** $Zn(s) \rightarrow Zn^{2+}(aq) + 2e^{-}$ oxidation
   When a metal goes into solution as ions, the electrons are left behind on the metal. This must be taking place to make the zinc negative.

b $Cu^{2+}(aq) + 2e^- \rightarrow Cu(s)$ reduction
When metal ions come out of solution and deposit themselves as neutral atoms, they take up electrons from the metal. This must be taking place to make the copper positive.

2. Half-reactions at the negative electrode:
Cell 2: $Cu(s) \rightarrow Cu^{2+}(aq) + 2e^-$
Cell 3: $Zn(s) \rightarrow Zn^{2+}(aq) + 2e^-$
Half-reactions at the positive electrode:
Cell 2: $2Ag^+(aq) + 2e^- \rightarrow 2Ag(s)$
Cell 3: $2Ag^+(aq) + 2e^- \rightarrow 2Ag(s)$
3. Cell 1: $Zn(s) + Cu^{2+}(aq) \rightarrow Zn^{2+}(aq) + Cu(s)$
Cell 2: $Cu(s) + 2Ag^+(aq) \rightarrow Cu^{2+}(aq) + 2Ag(s)$
Cell 3: $Zn(s) + 2Ag^+(aq) \rightarrow Zn^{2+}(aq) + 2Ag(s)$
4. The reactions are identical to those which occurred in Experiment 7.9.
5. The salt bridge completes the circuit by allowing ions carrying charge to move between the two half-cells. These positive and negative ions replace those which are discharged at the electrodes, but contamination of the half-cell solutions is kept to a minimum.

EXPERIMENT 7.11 Testing predictions about redox reactions

Your students should make predictions using standard electrode potential values ($E^\ominus$) **before** they do the experiment. Reaction F does not occur as expected, presumably because of a kinetic effect (high activation energy). This limitation on predictions from $E^\ominus$ values is well worth emphasising.

Specimen results

Results Table 7.8

Reaction	Additional test	Observations	Deductions
A	Add Volasil 244	The upper layer became purple	$I_2(aq)$ is present, so the reaction has taken place
B		No change	A reaction has not taken place
C	Add potassium hexacyano-ferrate(III)	A dark blue precipitate was formed	$Fe^{2+}(aq)$ is present, so the reaction has taken place
D	Add Volasil 244	The upper layer became orange-brown	$Br_2(aq)$ is present, so the reaction has taken place
E		The purple solution became pale blue	$MnO_4^-(aq)$ reduced to colourless $Mn^{2+}(aq)$. Cu(s) oxidised to blue $Cu^{2+}(aq)$
F		No change	A reaction has not taken place

Answer to question

Predictions using $E^\ominus$ values were borne out in practice for experiments A to E but **not** for F.

EXPERIMENT 7.12 Variation of cell e.m.f. with concentration

Since the Nernst equation is not covered by any current syllabus, this experiment has been made fairly simple and you may feel that it could be omitted. However, it is quite satisfying to be able to confirm that one's results are in accordance with Le Chatelier's principle.

The number of measurements has been limited because results are likely to be unreliable at concentrations of 0.01 M or less, particularly in the case of the zinc sulphate solutions.

Specimen results

Results Table 7.9

$[Zn^{2+}(aq)]$/mol dm^{-3}	$[Cu^{2+}(aq)]$/mol dm^{-3}	ΔE/V
1.0	1.0	+1.06
1.0	0.10	+1.03
0.10	1.0	+1.09

Answers to questions

1. Diluting the copper ion solution results in a decrease of e.m.f.
2. Diluting the zinc ion solution results in an increase of e.m.f.
3. The equilibrium at the copper electrode is:

$$Cu^{2+}(aq) + 2e^{-} \rightleftharpoons Cu(s)$$

Reducing $[Cu^{2+}(aq)]$ would result in the equilibrium shifting to the left, producing more electrons. This would make the electrode less positive (more negative) so the e.m.f. would be reduced.

By a similar argument to the above, the zinc electrode would become more negative, so the e.m.f. would be increased.

EXPERIMENT 7.13 Rusting of iron and steel

This experiment, although fairly simple, requires careful observation. It may therefore be suitable as a test of observational skills and so can be used in the scheme of practical coursework assessment. Specimen results are given below, together with a suggested marking scheme for observation and recording and for interpretation.

Suggested mark scheme

Total marks for observing and recording = 20, as follows:

Steel sheet	Observes dark blue ppt. at centre of drop	(2)
	Observes red solution near edge of drop	(2)
Nail A	Observes dark blue ppt., forming more rapidly at the ends	(2)
	Observes red solution between the blue areas	(2)
Nail B	Observes rapid formation of blue ppt., faster than for nail A	(2)
	Observes red solution around the copper	(2)
Nail C	Observes red solution around nail, but no blue ppt.	(2)
	Observes slow formation of white ppt. around the zinc	(2)
Nail D	Observes very little or no blue ppt. and no red solution	(2)
Nail E	Same as for nail D	(2)

Total mark	20–18	17–14	13–10	9–6	5–2	1–0
Scaled mark	5	4	3	2	1	0

Total marks for interpretation = 12 for answers to questions, as follows:

1. Explanation of blue ppt. + ion–electron half-equation (2)
   Explanation of red solution + ion–electron half-equation (2)
3. Iron, copper and solution form an electrochemical cell (1)
   Explanations of processes at iron and copper + equations (2)
4. Iron, zinc and solution form an electrochemical cell (1)
   Explanations of processes at iron and zinc + equations (2)
5. Iron phosphate formed on nail D (1)
   Iron phosphate and grease form protective layers (1)

Total mark	12–11	10–9	8–6	5–4	3–2	1–0
Scaled mark	5	4	3	2	1	0

Specimen results

Results Table 7.10

Experiment	Observations
Steel sheet	Fairly rapid appearance of dark blue ppt. in the centre of the drop. Gradual appearance of pink/red solution near the edge of the drop, all around
Nail A	Fairly rapid appearance of dark blue ppt. at several places along the nail, particularly at the ends. Red solution appears between the blue areas
Nail B (+Cu)	Rapid development of dark blue ppt. along most of the nail, faster than in the case of nail A. Red solution around the copper
Nail C (+Zn)	Red solution developing around the nail, but no blue ppt., even after 10 minutes. White ppt. slowly forms around the zinc
Nail D (H_3PO_4)	Very small amount of blue ppt. formed, much less than with nail A. Also little or no red solution
Nail E (grease)	Again, little or no blue ppt. formed. Much less that with nail A.

Answers to questions

1. The blue precipitate was ‘Prussian blue’, indicating the presence of Fe^{2+} ions. Therefore the iron corroded at the centre of the drop.
   $Fe(s) \rightarrow Fe^{2+}(aq) + 2e^-$
   The red solution near the outer edge of the drop was due to the reaction of phenolphthalein with hydroxide ions.
   $O_2(g) + 2H_2O(l) + 4e^- \rightarrow 4OH^-(aq)$
2. For nail A, corrosion occurred more rapidly at the ends.
3. Nail B corroded much faster than nail A. The copper, iron and the solution formed an electrochemical cell. The iron corroded and hydroxide ions were formed at the copper surface.
   Iron surface $Fe(s) \rightarrow Fe^{2+}(aq) + 2e^-$
   Copper surface $O_2(g) + 2H_2O(l) + 4e^- \rightarrow 4OH^-(aq)$
4. Nail C did not appear to corrode at all. The white precipitate near the zinc indicated that the zinc was corroding.
   $Zn(s) \rightarrow Zn^{2+}(aq) + 2e^-$
   This is because zinc has a much more negative electrode potential than iron. Hydroxide ions were formed at the iron surface.
   $O_2(g) + 2H_2O(l) + 4e^- \rightarrow 4OH^-(aq)$

5. The rates of corrosion of nails D and E were much slower than for nail A. This was because a protective film was formed on the iron surface. On nail D a coating of iron phosphate was formed.

EXPERIMENT 7.14 Determination of the Faraday constant

It is important to maintain a low current density. If it is too high, the deposited copper layer will be coarse and easily detached. For this reason also, the cathode is dried by careful rinsing with propanone and the plated area should not be touched.

Specimen results

Initial mass of copper cathode	=	38.420 g
Final mass of copper cathode	=	38.595 g
Current	=	0.20 A
Time	=	45 min.

Calculations

1. Mass of copper deposited = 38.595 g − 38.420 g = 0.175 g
2. Amount of copper deposited = mass of copper/molar mass of copper
   = 0.175 g/63.5 g mol^{-1} = 2.76×10^{-3} mol
   ∴ amount of electrons transferred = $2 \times 2.76 \times 10^{-3}$ mol = 5.52×10^{-3} mol
3. Quantity of electricity passed = current × time
   = 0.20 A × 45 × 60 s = 540 C
4. Faraday constant = quantity of electricity/amount of electrons
   = 540 C/(5.52×10^{-3} mol) = **97 800 C mol^{-1}**

EXPERIMENT 8.1 Chemical properties of ethanol

Please see the notes on pages 4 and 5, which give details of the hazards of using organic chemicals, and describe how to dispose of them safely.

To save time, you could allocate tests to different students and tell them to share their results. Students working alone could omit F, and B or C, and make use of the specimen results.

For Tollens test in part B, a silver mirror will form only on a clean test-tube. If the tube is greasy, a silver precipitate is formed instead.

For the Fehling's test in part B, ethanal tends to form a resin in the alkaline conditions, which may mask the formation of red copper(I) oxide. The students may therefore note a greenish-yellow rather than red colour, which is acceptable as a positive test for a reducing agent.

In part F (dehydration of ethanol) great care must be taken when heating the ethanol on ceramic wool to prevent the pressure of the vapour causing the pumice to block the tube. This experiment **must** be done in a fume cupboard with the front down as far as is practically possible. Students may not prepare enough ethene for both tests although, of course, each test may be done on half a tube of gas. Ethene should not be tested with bromine dissolved in an inert solvent because 1,2-dibromoethane is toxic. Bromine water is used instead.

Specimen results

Results Table 8.1
Reactions of ethanol

Property/Reaction	Observations
A **Solubility of water** pH of solution	Completely soluble pH 6 before and after addition
B **Mild oxidation** **a** Smell **b** Fehling's solution **c** Silver mirror (Tollens) test	**a** Pungent smell **b** Blue colour changed through green to reddish-brown precipitate **c** A silver mirror developed on the inside of the tube
C **Further oxidation** **a** Smell **b** Universal indicator paper **c** Sodium carbonate	**a** Vinegary smell **b** Orange; weak acid, pH 3–4 **c** Effervescence (CO_2)
D **Reaction with sodium**	Dissolved quietly; gas evolved popped with lighted splint (hydrogen); Evaporation gave white solid
E **Esterification**	Pleasant, sweet smell
F **Dehydration** **a** Bromine water **b** Acidified potassium manganate(VII) solution	**a** Decolorised **b** Decolorised

Answers to questions

1. Reaction B. The mild oxidation of ethanol produced a compound which acts as a reducing agent by giving positive Fehling's and silver mirror (Tollens) tests. The reducing agent is ethanal, CH_3CHO.
2. In these reactions, dichromate(VI) ions are reduced to chromium(III) ions.

$$Cr_2O_7^{2-}(aq) + 14H^+(aq) + 6e^- \rightarrow 2Cr^{3+}(aq) + 7H_2O(l)$$

dichromate(VI) (orange) → chromium(III) (blue–green) (violet if excess acid, no $Cr_2O_7^{2-}$)

3. Reaction C. Further oxidation of ethanol gave a product which turned universal indicator orange and is therefore acidic. The product smelt of vinegar – ethanoic acid.
4. **a** The use of a higher proportion of ethanol than acidified sodium dichromate and distilling off the product as fast as it is formed favours the formation of ethanal.
   b The use of a higher proportion of acidified sodium dichromate than ethanol and refluxing the mixture favours the formation of ethanoic acid.
5. Hydrogen gas was detected, indicating that CO—H fission must have taken place.
6. The fact that the product immediately decolorised bromine water and acidified potassium manganate(VII) indicates that a compound with a C = C bond (an alkene) was produced.
7. Reaction F produced a pleasant, sweet-smelling compound – an ester.

EXPERIMENT 8.2 Chemical properties of phenol

Please see the notes on pages 4 and 5, which give details of the hazards of using organic chemicals, and describe how to dispose of them safely.

The reaction between phenol and sodium can be violent and should be closely supervised (or demonstrated, if you wish). Students should be provided with very small pieces of sodium and not allowed to cut their own from stock. Note that phenol is not sufficiently acidic to react with sodium hydrogencarbonate. You might wish to use this experiment for assessment purposes.

Suggested mark scheme

Total marks for observing and recording = 10, as in Results Table 8.2 below.
Total marks for interpretation = 10, based on answers to questions 2–8, below.

Total mark	10–9	8–7	6–5	4–3	2–1	0
Scaled mark	5	4	3	2	1	0

Specimen results

Results Table 8.2
Reactions of phenol

Test	Observations
A **Solubility in water**	
a A little phenol	Slowly dissolved (1)
b A lot of phenol	No more dissolved (1)
c pH of solution	pH = 4 (1)
B **Reaction with sodium hydrogencarbonate**	No reaction (1)
C **Reaction with sodium hydroxide**	Phenol dissolved readily – far more soluble than in water (1)
Subsequent addition of hydrochloric acid	Phenol reappeared in an oily mixture (1)
D **Reaction with sodium**	Vigorous effervescence Gas ignited with a 'pop' (1)
E **Reaction with bromine water**	White precipitate (1) Brown colour disappeared (1)
F **Action of neutral iron(III) chloride solution**	Violet coloration (1)

Answers to questions

1. The smell of phenol often reminds people of antiseptics and/or disinfectants. It was formerly used in 'carbolic soap', a mild antiseptic, and some modern products contain similar compounds, e.g. TCP (trichlorophenol).
2. Phenol is not as soluble as ethanol; whereas ethanol is neutral, phenol is acidic. (1)
3. The fact that hydrogen is produced indicates the O—H bond has been broken in phenol, as it is in ethanol. (1)
4. **a** Reaction with neutral iron(III) chloride – phenol gives a violet colour, ethanol does not.
   b Reaction with bromine water – phenol gives a white precipitate, ethanol does not.
   c pH of solution – phenol gives an acidic solution in water, ethanol does not. (3)
5. Phenol dissociates to a certain extent in water, as indicated by the fact that a solution of it in water turns universal indicator orange/red. However, with the addition of sodium hydroxide, there is interaction between the hydroxide ions and hydronium ions. This shifts the equilibrium to the right promoting the formation of phenoxide ions.
   $C_6H_5OH + H_2O \rightleftharpoons C_6H_5O^- + H_3O^+$ (2)

6. Test A – pH of solution. Ethanol is not sufficiently acidic to donate protons to water and therefore appears neutral in solution. (1)
7. Test B – reaction with sodium hydrogencarbonate. Unlike dilute mineral acids, phenol is not sufficiently acidic to liberate carbon dioxide from a hydrogencarbonate. (1)
8. Substitution must occur in the benzene ring, and appears to be made easier by the presence of the —OH group. (1)

EXPERIMENT 8.3 Reactions of amines

Please see the notes on pages 4 and 5, which give details of the hazards of using organic chemicals, and describe how to dispose of them safely.

This experiment compares the reactions of phenylamine and butylamine with those of ammonia. The reaction with hydrochloric acid gives a salt in each case; phenylammonium chloride is immediately visible as a solid, whereas butylammonium chloride and ammonium chloride are only visible after evaporation of water.

The instructions for the reactions with nitrous acid are necessarily rather complicated; we suggest that you make yourself familiar with the details so that you can help students to follow them and understand what they are doing. Note: We have preferred naphthalen-2-ol to naphthalen-1-ol because it is less toxic.

Specimen results

Results Table 8.3

Test	Phenylamine	Butylamine	Ammonia
A **a** Solubility in water	Apparently insoluble	Soluble	Very soluble
b pH of solution	6–7 (same as water)	11–14	11–14
B **a** Reaction with hydrochloric acid	Dense white fumes; clear solution; exothermic	Dense white fumes; clear solution; exothermic	White fumes; clear solution; exothermic
b Evaporation of water from product	Colourless crystals immediately	Colourless crystals	Colourless crystals
C Reaction with copper(II) sulphate solution	Bright green colour which first darkens, then gives a yellow–green precipitate	Blue–green precipitate which darkens and then dissolves to a deep blue solution	Pale blue precipitate which darkens and then dissolves to a deep blue solution
D **a** Reaction with nitrous acid below 5 °C	No apparent reaction apart from slight decomposition of nitrous acid	Slow bubbling; gas has no effect on lime-water and puts out a lighted splint	Slow bubbling; gas has no effect on lime-water and puts out a lighted splint
b Reaction of product from **a** with:			
i) phenol	Bright yellow precipitate	Yellowish solution	No change
ii) naphthalen-2-ol	Scarlet red precipitate	Yellowish solution	No change
c Effect of heat on product from **a**	Fast bubbling; gas has no effect on lime-water and puts out a lighted splint	Fast bubbling; gas has no effect on lime-water and puts out a lighted splint	Fast bubbling; gas has no effect on lime-water and puts out a lighted splint
E Reaction with bromine water	Brown colour disappears; white 'woolly' precipitate	No change	No change

Answers to questions

1. Butylamine is at least as basic as ammonia, whereas phenylamine is weaker.
2. Butylamine is soluble in water due to the polar nature of the N—H bond, which leads to hydrogen bonding with water and favours dissociation. In phenylamine, the non-polar benzene ring diminishes the influence of the polar NH_2 group and the

position of equilibrium for the dissociation of phenylamine in water is well over to the left.

$$RNH_2 + H_2O \rightleftharpoons RNH_3^+ + OH^-$$

However, phenylamine is very soluble in hydrochloric acid. Since this is a stronger acid than water, the position of equilibrium is well over to the right.

$$C_6H_5NH_2 + HCl \rightleftharpoons C_6H_5NH_3^+ + Cl^-$$

3. An arylamine reacts with bromine water to give a white precipitate whereas an alkylamine does not.
   An arylamine reacts with cold nitrous acid and an alkaline solution of naphthalen-2-ol to give a red dye whereas an alkylamine does not.

EXPERIMENT 8.4 Reactions of aldehydes and ketones

Please see the notes on pages 4 and 5, which give details of the hazards of using organic chemicals, and describe how to dispose of them safely.

In this experiment the students perform simple test-tube reactions on ethanal and propanone. From their observations they should deduce that, although there are similarities in the reactions of aldehydes and ketones, there are some distinct differences.

The reaction with 2,4-dinitrophenylhydrazine may fail to give a precipitate if too much aldehyde or ketone is used; the hydrazone may dissolve in excess carbonyl compound.

In Fehling's test, ethanal tends to form a resin in the alkaline conditions, which may mask the formation of red copper(I) oxide. The students may therefore note a greenish-yellow (rather than a red) colour which is acceptable as a positive test for a reducing agent.

In Tollens test, a silver mirror will form only in a clean test-tube. Also, it is important that only sufficient ammonia is added to barely dissolve the silver oxide – some black specks of silver oxide should still remain.

Specimen results

Results Table 8.4

Test	**Observations**	
	Ethanal	Propanone
A Condensation reaction with 2,4-dinitrophenylhydrazine	Orange–yellow precipitate	Orange–yellow precipitate
B Oxidation reactions:		
a with acidified dichromate(VI)	Orange solution turned blue–green	No change
b with Fehling's solution	Blue solution became cloudy green, then formed a reddish precipitate	No change
c with Tollens reagent	Grey precipitate*	No change
C Triiodomethane reaction	Pale yellow precipitate	Pale yellow precipitate

*Silver is usually produced too rapidly to form a good silver mirror.

Answers to questions

1. The oxidation reactions with acidified dichromate(VI), Fehling's solution and Tollens reagent, all serve to distinguish between ethanal and propanone.
2. 2,4-dinitrophenylhydrazine could be used in a general test for carbonyl compounds since both aldehydes and ketones react with it to produce orange precipitates.

EXPERIMENT 8.5 Identifying an unknown carbonyl compound

Please see the notes on pages 4 and 5, which give details of the hazards of using organic chemicals, and describe how to dispose of them safely.

The students should be provided with an unknown aldehyde or ketone (compound X), selected from the following list (preferably not the most familiar ones, methanal, ethanal and propanone):

■ methanal (formaldehyde)	$HCHO$
■ ethanal (acetaldehyde)	CH_3CHO
■ propanal (propionaldehyde)	C_2H_5CHO
■ butanal (n-butyraldehyde)	C_3H_7CHO
■ 2-methylpropanal (isobutyraldehyde)	$(CH_3)_2CHCHO$
■ benzaldehyde	C_6H_5CHO
■ propanone (acetone)	CH_3COCH_3
■ butanone (methyl ethyl ketone)	$CH_3COC_2H_5$
■ pentan-2-one (methyl n-propylketone)	$CH_3COC_3H_7$
■ pentan-3-one (diethyl ketone)	$C_2H_5COC_2H_5$
■ hexan-2-one (methyl n-butyl ketone)	$CH_3COC_4H_9$
■ 4-methylpentan-2-one (methyl isobutyl ketone)	$CH_3COCH_2CH(CH_3)_2$
■ cyclohexanone	$C_6H_{10}O$

In the first part of the experiment, the students identify the unknown compound as either an aldehyde or a ketone. In the second part, they prepare a derivative and determine its melting-point. The table of melting points of 2,4-dinitrophenylhydrazones printed in *Advanced Practical Chemistry* (Table 8.1) should enable them to identify the compound. They should check with you or your technician whether they have identified X correctly.

If the ILPAC video programme 'Organic techniques II' is not available for your students, we suggest that you demonstrate the procedures for recrystallisation and melting point determination.

Answers to questions

1. The solvent must dissolve the crude material readily when hot, but only to a small extent when cold. Ethanol fits both of these requirements for this particular experiment.
2. When the derivative crystallises, soluble impurities remain in solution and are separated from the derivatives by filtration.
3. Some of the derivative is always lost during filtering because the filtrate is a saturated solution. Using the minimum amount of solvent reduces this loss.
4. Insoluble impurities may be removed by filtering the hot solution before crystallisation occurs. (The funnel and flask must be heated first, by pouring hot solvent through it, in order to prevent crystallisation occurring in the apparatus.)
5. Isomeric aldehydes and ketones frequently have closely similar boiling points, but the melting points of their derivatives usually differ considerably. Furthermore, boiling points are less readily measured with precision than melting points, and it is generally more difficult to obtain liquids in a high state of purity.

EXPERIMENT 8.6 Reactions of carbohydrates

Please see the notes on pages 4 and 5, which give details of the hazards of using organic chemicals, and describe how to dispose of them safely.

This experiment is divided into five parts, which can be completed in a 90-minute practical session. If your students do the polarimetry section, you may need to modify our instructions to suit your polarimeter.

The inversion of sucrose can be used as an experiment in reaction kinetics if you wish. In normal conditions, the concentrations of water and of hydrogen ions are constant, so that the reaction is first order with respect to sucrose. Since $(\alpha_t - \alpha_\infty) \propto$ [sucrose(aq)], students could measure the slope of their curve at a number of points to obtain a series of values of rate of reaction and [sucrose(aq)], which give a straight line when plotted against one another.

The mono-derivatives with 2,4-dinitrophenylhydrazine are usually slow to form because they are fairly soluble in water – you may feel that this part of the experiment could be omitted.

Specimen results

Results Table 8.5

Experiments	Observations				
	Glucose	**Fructose**	**Sucrose**	**Maltose**	**Starch**
A **Dehydration**					
a Action of heat	Melted easily to a clear liquid which became dark and viscous. Steamy gas		Melted easily to a clear liquid which became dark and viscous. Steamy gas		
b Sulphuric acid	Mixture turned black and swelled giving off heat and much steam. Crumbly black solid left. CO_2 and SO_2 detected		Mixture turned black and swelled giving off heat and much steam. Crumbly black solid left. CO_2 and SO_2 detected		
B **Oxidation**					
a Fehling's solution	Reddish ppt.	Reddish ppt.	No change	Reddish ppt.	
b Tollens reagent	Silver mirror	Silver mirror	No change	Silver mirror	
C **Carbonyl derivatives** 2,4-dinitro-phenylhydrazine	Yellow ppt. (slow)	Yellow ppt. (slow)			
D **Hydrolysis**			Red ppt. with Fehling's solution or silver mirror with Tollens reagent		B and C. Hydrolysis to reducing sugar

Results Table 8.6

	Movement of analyser from zero								
Glucose	Initial +27°; $\frac{1}{2}$ volume +13°; Diluted +13°	+ indicates clockwise; − indicates anticlockwise							
Sucrose	Time, t/min	0	3	6	10	15	20	30	∞
	Analyser reading, α_t/°	+34	+18	+7	−1	−8	−9	−11	−12
	$\alpha_t - \alpha_\infty$/°	46	30	19	11	4	3	1	0

Answers to questions

1. The presence of —OH groups gives rise to hydrogen bonding with water molecules.
2. The main products are carbon and water:

$$C_6H_{12}O_6 \rightarrow 6C + 6H_2O$$

3. The main reaction is called dehydration because water molecules are removed.
4. Some of the sulphuric acid is reduced by the hot carbon to sulphur dioxide, while some of the carbon is oxidised to carbon dioxide:

$$C(s) + 2H_2SO_4(l) \rightarrow CO_2(g) + 2SO_2(g) + 2H_2O(l)$$

Some is merely diluted by the water released.

5. No, we only expected the glucose as an aldehyde to give positive Tollens and Fehling's tests. We were surprised to find that fructose as a ketone also gave positive results in these tests. The reason is that even though fructose has no —CHO group, the combination of C=O and the neighbouring C—OH group is an effective reductant.
6. In each case glucose is oxidised to gluconic acid (the CHO group becomes —CO_2H). Tollens reagent is reduced to metallic silver, and Fehling's solution is reduced to copper(I) oxide.
7.

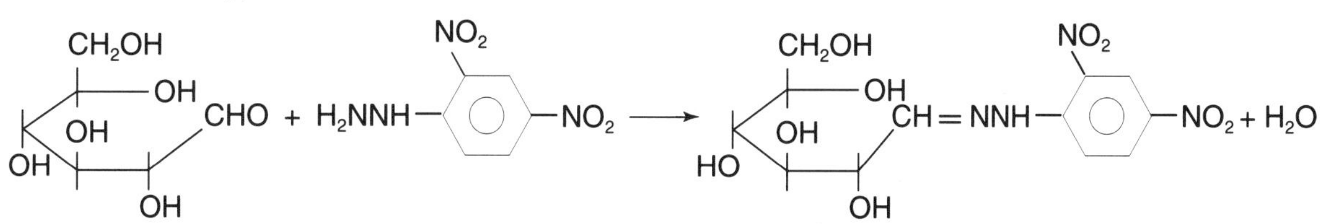

8.

$$\underset{\text{sucrose}}{C_{12}H_{22}O_{11}} + H_2O \rightarrow \underset{\text{glucose}}{C_6H_{12}O_6} + \underset{\text{fructose}}{C_6H_{12}O_6}$$

9. Both sucrose and glucose are dextrorotatory. Fructose is laevorotatory to a greater extent than glucose is dextrorotatory. The final mixture is therefore laevorotatory. 'Inversion' refers to the change from dextrorotation to laevorotation, (+) to (−).
10. Biological catalysts called enzymes secreted by the human digestive system enable hydrolysis to take place at body temperature.
11. When starch was warmed in the presence of the enzyme amylase (present in human saliva) as in tube B, it was hydrolysed to the disaccharide maltose which has reducing properties.

When the starch was boiled with acid as in tube C, it was hydrolysed to glucose. The starch remained unchanged in tube A, since the enzyme had been destroyed (denatured) by heat.

Tube D was the control and the starch remained unchanged.

12. In sucrose and starch the link between the sugar units is via the aldehyde groups which then become ineffective. After hydrolysis these aldehyde groups are free to give a positive Fehling's test.
13. Before the body can utilise the energy from carbohydrates they must be broken down by hydrolysis into monosaccharides such as glucose which can then be transported around the body.
14. The addition of iodine solution to starch produces a blue–black colour.
15.

$$[\alpha] = \frac{\alpha}{lc} \therefore c = \frac{\alpha}{[\alpha]l} = \frac{27^\circ}{52.5^\circ\,g^{-1}\,cm^3\,dm^{-1} \times 1\,dm} = \mathbf{0.51\,g\,cm^{-3}}$$

16. $\alpha_t - \alpha_\infty$ is proportional to the amount of sucrose present in the solution. The slope of the curve is therefore the rate of reaction and this decreases to zero as the reaction proceeds to completion. This is a good example of a first order reaction.

EXPERIMENT 8.7 Observation and deduction exercise 3

Please see the notes on pages 4 and 5, which give details of the hazards of using organic chemicals, and describe how to dispose of them safely.

This experiment is intended primarily for students taking an A-level practical examination in which one of the questions involves the identification of an unknown compound, or for students being internally assessed on their practical work.

For those using this as a practical test, advise your students to study the wording of the question closely since the type of answer required varies from paper to paper. For example, sometimes details of additional tests must be given and sometimes equations and/or comments on the type of reactions are required.

Point out to your students that a high proportion of marks is allocated to observations, even when no inferences are made.

C = Glacial ethanoic acid, CH_3COOH
D = Ethanol, CH_3CH_2OH
E = Phenylammonium chloride, $C_6H_5NH_3^+Cl^-$

Suggested mark scheme

Total marks for observing and recording = 18.
Record marks from observations in Results Table 8.7.

Total mark	18–16	15–12	11–9	8–6	5–3	2–0
Scaled mark	5	4	3	2	1	0

Total marks for interpretation (inferences) = 17.
Record marks from inferences in Results Table 8.7 and from Results Table 8.8 (together with students' suggestions for structural features of E, which follow it).

Total	17–14	13–10	9–7	6–5	4–3	2–0
Scaled mark	5	4	3	2	1	0

Specimen results

Results Table 8.7

Test	Observations	Inferences
1. **a** Place 1 cm^3 of C in a test-tube and add an equal volume of water. Now add a little anhydrous sodium carbonate **b** Repeat test 1**a** using D	Effervescence (1) Gas turned lime-water milky (1) No reaction (1)	CO_2 produced which suggests C is a carboxylic acid (1) D is not a carboxylic acid (1)
2. (The reactions 2**a** and 2**b** should be performed at a fume cupboard) **a** Place 2 or 3 cm^3 of C in a dry beaker. Add a little phosphorus pentachloride (**care**) **b** Repeat test 2**a** using D (**care**)	Vigorous reaction giving a steamy gas (1) Gas gave white fumes with ammonia (1) Vigorous reaction giving a steamy gas (1) Gas gave white fumes with ammonia (1)	HCl produced, which suggests C contains an —OH group (1) HCl produced, which suggests D contains an —OH group (1)
3. Mix, in a small beaker, about 2 cm^3 of each of C, D and concentrated sulphuric acid (**care**). Warm gently but do not boil. Pour the mixture into an excess of aqueous sodium carbonate in a large beaker. Smell the product	A sweet, fruity smell was noticed (1)	An ester was probably formed from a carboxylic acid, C, and an alcohol, D (1)
4. Mix about 5 cm^3 of D with an equal volume of aqueous potassium dichromate in a test-tube. Pour the mixture into about 10 cm^3 of nearly boiling dilute sulphuric acid in a small beaker. Smell the mixture	A pungent smell (1) The solution changed from orange to blue–green (1)	The dichromate has oxidised D (suspected alcohol) to a pungent smelling aldehyde (1)
5. In a small beaker dissolve a few crystals of potassium iodide in about 10 cm^3 of D. Add a few drops of aqueous sodium hydroxide. Now add 2 or 3 cm^3 of aqueous sodium chlorate(I) (hypochlorite). Warm the mixture but do not boil. Cautiously smell the products	Yellow precipitate (1) 'Antiseptic' smell (1)	Triiodomethane produced, which suggests that D contains the group $CH_3CH(OH)$ (1)
6. Make a solution of E in distilled water and use portions for the following tests: **a** To 2–3 cm^3 of the solution add aqueous bromine **b** To 2–3 cm^3 of the solution add aqueous silver nitrate. Then add dilute nitric acid. **c** Finally add dilute aqueous ammonia	Bromine decolourised (1) White precipitate formed (1) White precipitate (1) Insoluble in dilute nitric acid (1) but soluble in dilute ammonia (1)	This suggests that E is either an aromatic amine or a phenol (1) Precipitate of AgCl suggests E contains Cl^- or a readily hydrolysed —Cl group (1)
7. Dissolve a little E in about 1 cm^3 of concentrated hydrochloric acid and dilute to about 4 cm^3 with distilled water. Cool the tube in an ice–water mixture and add a few drops of aqueous sodium nitrite (to be prepared by dissolving sodium nitrite in distilled water). Leave the tube in the ice–water mixture. Dissolve a few crystals of phenol (**care**) in 7–8 cm^3 of aqueous sodium hydroxide, cool this solution, and then add it to the cold solution prepared as above	A dullish orange precipitate was formed (1)	An azo-dye was probably formed from a primary aromatic amine (1) This and the previous tests suggest that E is a salt of a primary aromatic amine, $ArNH_3{}^+Cl^-$ (1)

Results Table 8.8

	Reasoning
Functional group in C —OH (1)	Reacts with PCl_5 to evolve HCl, suggesting the presence of an —OH group. Since it fails to evolve CO_2 from Na_2CO_3, it must be the —OH group of an alcohol (1)
Functional group in D (1) 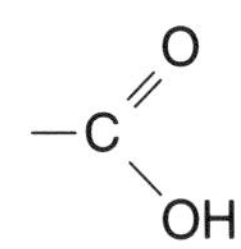	Reacts with PCl_5 to evolve HCl, suggesting an —OH group. Since it evolves CO_2 from Na_2CO_3, it must be the —OH group of a carboxylic acid (1)

The structural features of compound E are as follows. E contains at least one benzene ring, to which is attached an $NH_3^+Cl^-$ group. (2)

(Practical test total: 35 marks)

EXPERIMENT 9.1 Determining the molar mass of a gas

You may prefer to use a syringe method, but we have found that students do not readily understand the buoyancy correction and, in any case, get poorer results using a syringe.

If you allow students to operate a gas cylinder, be sure to instruct them carefully – we suggest that they use a needle valve **only**, after you have opened the main valve and set the regulator.

If you use a gas generator, for example, Kipp's, you should instruct the students (and/or technician) how to purify the gas before use.

Some students may realise that the densities given refer to **dry** air. However, this makes little difference to the results, which can be very good with this simple method.

(The treatment of results does not explore the effect of Archimedean upthrust on the empty flask, as students appear to find this distracting. In fact, the calculation implies that the volume of air inside the flask is equal to that of the air displaced by the flask, a not unreasonable assumption.)

Specimen results and calculation

Results Table 9.1

Mass of flask filled with air	47.933 g
Mass of flask filled with CO_2	47.998 g
Mass of flask filled with water	152.8 g
Room temperature	26 °C
Atmospheric pressure	757 mmHg
Density of air under condition of experiment	0.001 18 g cm^{-3}

Volume of CO_2 = volume of flask

$$= \frac{\text{mass of water}}{\text{density of water}} = \frac{152.8\text{ g} - 47.9\text{ g}}{1.00\text{ g cm}^{-3}} = 104.9\text{ cm}^3$$

Mass of empty flask = mass of flask and air − mass of air
= 47.933 g − (volume of air × density of air)
= 47.933 g − ($104.9\ cm^3 \times 0.001\,18\ g\ cm^{-3}$)
= 47.933 g − 0.124 g = 47.809 g

Mass of CO_2 = mass of flask and CO_2 − mass of flask
= 47.998 g − 47.809 g = 0.189 g

$$pV = nRT = \frac{mRT}{M}$$

$$\therefore\ M = \frac{mRT}{pV}$$

If we use $R = 0.0821\ atm\ dm^3\ K^{-1}\ mol^{-1}$,

then p must be in atm, i.e. $\frac{757\ mmHg}{760\ mmHg\ atm^{-1}} = 0.996\ atm$

and V must be in dm^3, i.e. $0.105\ dm^3$
and T must be in K, i.e. (273 + 26) K = 299 K

$$\therefore\ M = \frac{0.189\ g \times 0.0821\ atm\ dm^3\ K^{-1}\ mol^{-1} \times 299\ K}{0.996\ atm \times 0.105\ dm^3} = \mathbf{44.4\ g\ mol^{-1}}$$

This agrees well with the accepted value $44.1\ g\ mol^{-1}$. Students' results should lie between 43 and 46 $g\ mol^{-1}$.

Answers to questions

1. If the molar mass is $44.1\ g\ mol^{-1}$, the relative molecular mass is 44.1. (This question was included simply to remind students of the difference between the two terms.)

2. Density, $\rho = \frac{\text{mass of sample}}{\text{volume of sample}} = \frac{\text{molar mass}}{\text{molar volume}} = \frac{M}{V_m}$

$$\rho = \frac{44.1\ g\ mol^{-1}}{22.4\ dm^3\ mol^{-1}} = 1.97\ g\ dm^{-3} \text{ or } \mathbf{1.97 \times 10^{-3}\ g\ cm^{-3}}$$

3. Rapid removal of the delivery tube would cause turbulence and tend to reduce the pressure momentarily. This would encourage air to enter the flask.
4. The purpose of this weighing is to obtain the mass of water in the flask, which is over 100 g. An error of 0.1 g in 100 g is only 0.1%. The weighings on the more accurate balance are to obtain the mass of gas, which is only about 0.2 g. An error of 0.001 g in 0.2 g is 0.5% which is in fact **more** serious than the error in using the less accurate balance.

EXPERIMENT 9.2 Determining the molar mass of a volatile liquid

This experiment also gives good results but requires more manipulative skill, especially in transferring liquid to the gas syringe without loss.

Make sure that steam generators are fitted with safety tubes and that students connect them properly to avoid scalding.

Specimen results

Results Table 9.2

Mass of hypodermic syringe and liquid before injection	12.898 g
Mass of hypodermic syringe and liquid after injection	12.716 g
Temperature of vapour	100 °C
Atmospheric pressure	756 mmHg
Volume of air in syringe	5.0 cm^3
Volume of air and vapour in syringe	67.0 cm^3

Calculation

$$pV = nRT = \frac{m}{M}RT$$

$$\therefore M = \frac{mRT}{pV}$$

If we use $R = 0.0821$ atm dm^3 K^{-1} mol^{-1},

p must be in atm, i.e. $\frac{756}{760} = 0.995$ atm

V must be in dm^3, i.e. $\frac{67.0\ cm^3 - 5.0\ cm^3}{1000\ cm^3\ dm^{-3}} = 0.0620\ dm^3$

T must be in K, i.e. (273 + 100) K = 373 K

Then $M = \frac{(12.898\ g - 12.716\ g) \times 0.0821\ atm\ dm^3\ K^{-1}\ mol^{-1} \times 373\ K}{0.995\ atm \times 0.0620\ dm^3} = \mathbf{90.3\ g\ mol^{-1}}$

This agrees well with the accepted value 88.1 g mol^{-1}. Although this experiment is not quite as accurate as Experiment 9.1, Students' results should lie between 85 and 95 g mol^{-1}.

Answers to questions

1. Again, the relative molecular mass is found from the molar mass by dropping the units, i.e. 90.3.
2. If the needle were too short, some of the liquid injected might not reach the hottest part of the syringe and therefore might not vaporise. In this case, the volume of vapour would be too small, and the final value for M would be too large:

   $$M = \frac{mRT}{pV}$$

3. Volume of vapour at 756 mmHg and 373 K = 62.0 cm^3
   Therefore using the combined gas law:

   $$\text{volume at s.t.p.} = 62.0\ cm^3 \times \frac{756\ mmHg}{760\ mmHg} \times \frac{273\ K}{373\ K}$$

   $$= 45.1\ cm^3 = 0.0451\ dm^3$$

   Scaling up to 22.4 dm^3 requires a scale factor of $\frac{22.4}{0.0451}$

   $\therefore$ 22.4 dm^3 of vapour has a mass of $0.182\ g \times \frac{22.4}{0.0451} = \mathbf{90.4\ g}$

EXPERIMENT 9.3 Determining the molar mass of a gas by effusion

This experiment gives good reproducible results if the apparatus is set up carefully. You may find that students have wrist-watches with stop-watch facilities more accurate than laboratory stop-clocks – reading to 0.1 second appears to be justified.

If you allow students to operate a gas cylinder, be sure to instruct them carefully and supervise closely – we suggest that they use a needle valve **only**, after you have opened the main valve and set the regulator. Also ensure that no flames are in the laboratory.

During the course of the experiment, you might check that students understand the difference between diffusion and effusion, and that Graham's law applies to both. We have assumed some knowledge of simple diffusion experiments.

We do not think it necessary to explain why the pin-hole should be small, other than to obtain measurable rates of effusion, but you may wish to discuss this with more able students.

Specimen results and calculations

Results Table 9.3

	1	2	3	4	Mean
Time for effusion of 50 cm^3 of hydrogen/s	10.7	10.9	10.7	10.7	10.7
Time for effusion of 50 cm^3 of domestic gas/s	31.3	30.9	31.0	32.1	31.3

Graham's law states:

$$\frac{\text{rate of effusion of } H_2}{\text{rate of effusion of domestic gas}} = \sqrt{\frac{\text{density of domestic gas}}{\text{density of } H_2}}$$

Since density $\propto M$ and rate of effusion $\propto \frac{1}{t}$ for a fixed amount, we write:

$$\frac{t_{gas}}{t_{H_2}} = \sqrt{\frac{M_{gas}}{M_{H_2}}}$$

$$\therefore\ M_{gas} = M_{H_2} \times \left(\frac{t_{gas}}{t_{H_2}}\right)^2$$

$$= 2.02 \text{ g mol}^{-1} \times \left(\frac{31.3 \text{ s}}{10.7 \text{ s}}\right)^2 = \mathbf{17.3 \text{ g mol}^{-1}}$$

Students' results should lie between 16.0 g mol^{-1} and 19.0 g mol^{-1}.

Answers to questions

1. The main constituent of domestic gas is methane, CH_4, which has a molar mass of 16.0 g mol^{-1}. This is consistent with results for this experiment.
2. This experiment, carefully done, nearly always gives a result greater than 16.0 g mol^{-1}. This suggests that the major impurities have molar masses greater than 16.0 g mol^{-1}. The chemical composition of domestic gas varies a little but most supplies contain well over 90 per cent methane. The main impurities are nitrogen, and hydrocarbons such as ethane, C_2H_6, all of which have molar masses greater than 16.0 g mol^{-1}.
3. The piston should be allowed to fall from 75 cm^3 to 60 cm^3 before starting the stop-clock to ensure that air is displaced from the tap.
4. At a higher temperature, gas molecules move more rapidly and effusion time would therefore be shorter, but in the same ratio, so that the same value for molar mass should be obtained.
5. If the hole is larger, the rate of effusion would be larger and the times shorter. Provided the hole is not so large that times are difficult to record accurately, the experiment should give the same result. Note, however, that as the hole becomes larger, Graham's law holds less well.

EXPERIMENT 9.4 Investigating the hydrolysis of benzenediazonium chloride

This experiment is more straightforward than it appears at first sight. Make sure that your students do not start to collect the gas until the reaction mixture has reached a steady temperature, and that they understand why it does not affect the results to allow gas to escape before this point is reached. Students may also need help with the use of $V_\infty - V_t$ as a measure of the concentration of benzenediazonium chloride.

Better results are obtained if the reaction mixture is **continuously** stirred, but this is not practicable in most school laboratories. **Occasional** stirring should be avoided, since this results in uneven evolution of gas.

Check that the syringe piston moves freely and encourage your students to rotate it in order to equalise pressures before taking a volume reading. Also, make sure that they know in advance what to do when the gas volume exceeds the volume of the syringe.

Specimen results and calculations

Results Table 9.4

Time, t/min	0	1	2	3	4	5	6
Volume of N_2, V_t/cm^3	0	14	28	41	54	65	76
($V_\infty - V_t$)/cm^3	219	205	191	178	165	154	143
Time, t/min	7	8	9	10	11	12	13
Volume of N_2, V_t/cm^3	87	96	104	112	120	127	133
($V_\infty - V_t$)/cm^3	132	123	115	107	99	92	86
Time, t/min	14	15	16	17	18	19	20
Volume of N_2, V_t/cm^3	139	145	150	155	160	164	168
($V_\infty - V_t$)/cm^3	80	74	69	64	59	55	51
Time, t/min	21	22	23	24	25	∞	
Volume of N_2, V_t/cm^3	172	175	177	181	184	219	
($V_\infty - V_t$)/cm^3	47	44	42	38	35	0	

Temperature of thermostat: 47 °C

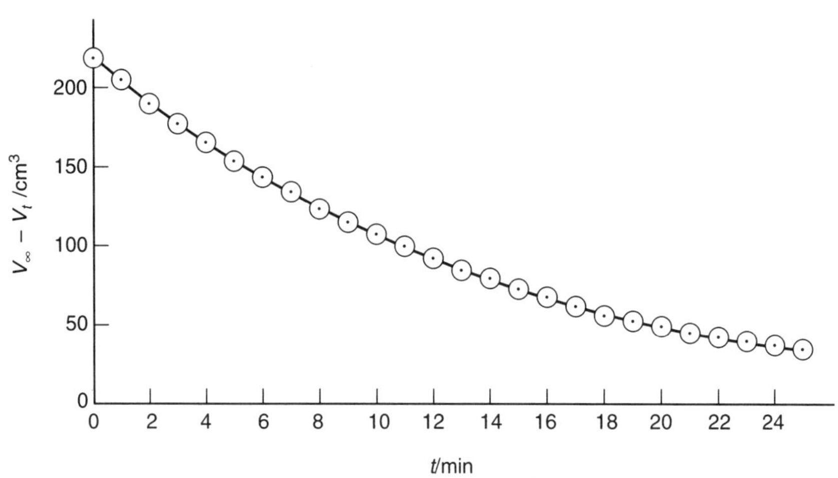

Results Table 9.5

Time /min	**Slope /cm^3 min^{-1}**	**Rate /cm^3 min^{-1}**	**$(V_\infty - V_t)$ /cm^3**
0	$\frac{219}{13.3}$	16.5	219
4	$\frac{214}{17.7}$	12.1	165
7	$\frac{202}{20.1}$	10.0	132
14	$\frac{150}{24.1}$	6.22	80
21	$\frac{88}{22.9}$	3.84	47

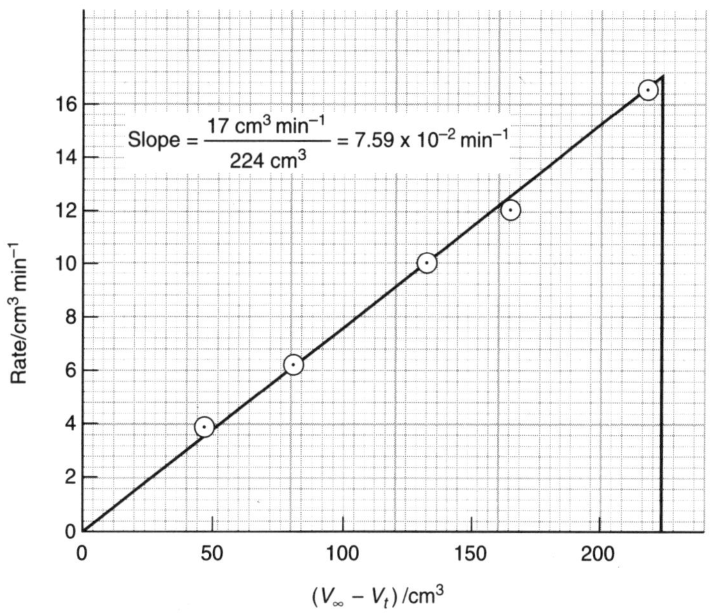

Answers to questions

1. Rate = $k[C_6H_5N_2^+Cl^-]$
2. $k = 7.59 \times 10^{-2}$ min^{-1} at 47 °C.
3. The initial gas escape can be neglected because we are interested in the volume change, $(V_\infty - V_t)$, e.g. if you had carried out the experiment and found that $V_\infty = 150$ cm^3 and $V_t = 50$ cm^3 then $(V_\infty - V_t) = 100$ cm^3.
   Suppose the volume of gas produced in the first four minutes was 10 cm^3, then $V_\infty = 160$ cm^3 and $V_t = 60$ cm^3 but $(V_\infty - V_t)$ is still 100 cm^3. Thus we can neglect the gas produced in the first four minutes.

EXPERIMENT 9.5 The kinetics of the reaction between iodine and propanone in aqueous solution

This experiment often causes problems because of the unreliability of inexpensive colorimeters. We suggest that you try out the procedure in advance in order to decide whether the variable light output of the lamp is more or less important than the precise positioning of the tube (see also our answer to question 5 on page 83). Optically matched tubes are preferable, but adequate results may be obtained with a carefully selected set of ordinary tubes, or even by inserting the same tube into the colorimeter for all readings. All tubes must be wiped clean at the lower end and handled only at the top.

Another problem is that heat from the bulb may increase the temperature of the mixture by several degrees over six minutes. However, this is not as serious as it seems, because it is the **initial** rate that is required. Calculations show that, for the initial concentrations we specify, each graph should be a virtually straight line; a temperature rise gives a curve showing the rate **increasing** slightly (as in our specimen results), but this has the same slope at $t = 0$. Note that the autocatalytic effect of increasing $[H^+(aq)]$ is not evident under these conditions because $[CH_3COCH_3(aq)]$ decreases as $[H^+(aq)]$ increases, and neither of these concentrations changes very much during the experiment.

Our results were obtained using a WPA CO65 colorimeter. Other colorimeters may give better results over a different range of iodine concentrations. If, for instance, a calibration curve suggests that the best range is from 0 to 0.005 mol dm^{-3}, you could use 0.010 M I_2 instead of 0.020 M I_2. Nothing else need be changed, because the rate of reaction is independent of the concentration of iodine.

Students sometimes obtain a series of values of $[I_2(aq)]$ which give a good line only if the calculated value at $t = 0$ is ignored. This is probably due to a 'zero drift' on the instrument. In this event, the slope of the line at $t = 30$ seconds should be taken.

If you are not satisfied that your colorimeter(s) can give satisfactory results, your students could study the same reaction by a titration method (see, for instance, *Nuffield Advanced Chemistry* (Third Edition), Topic 8) or they could study a different reaction by a 'clock' method (see Experiment 9.8).

Alternatively, you could try capturing the output from the colorimeter 'live' on a BBC microcomputer via the analogue port. For details see the articles by J.C.H. Stephens, *School Science Review*, 1988, **69**, (249), pages 762–7 and *School Science Review*, 1989, **71**, (254), pages 92–6.

Specimen results

Results Table 9.6

Volume of 0.02 M I_2 solution/cm^3	Volume of distilled water/cm^3	$[I_2(aq)]$ /10^{-3} mol dm^{-3}	Meter reading (% transmission)
0.0	10.0	0.0	100.0
1.0	9.0	2.0	79.5
2.0	8.0	4.0	61.0
3.0	7.0	6.0	51.0
4.0	6.0	8.0	42.0
5.0	5.0	10.0	37.0

For an outline calibration curve see page 82.

Results Table 9.7

***t* = time/min   % = % transmission   I = $[I_2(aq)]/10^{-3}$ mol dm^{-3}**

	t	0	½	1	1½	2	2½	3	3½	4	4½	5	5½	6
a	%	–	62.5	63.8	64.3	65.0	66.0	66.8	67.8	68.8	69.8	71.0	72.0	73.0
	I	4.00	3.90	3.72	3.66	3.56	3.44	3.32	3.20	3.07	2.94	2.80	2.70	2.58
b	%	–	62.5	64.0	65.8	67.8	69.8	71.8	74.2	76.5	79.8	82.5	85.8	89.0
	I	4.00	3.90	3.70	3.46	3.20	2.94	2.72	2.42	2.18	1.83	1.55	1.25	0.95
c	%	–	65.0	67.0	69.8	72.5	76.0	79.8	83.8	88.0	93.0	97.5	–	–
	I	4.00	3.55	3.30	2.94	2.62	2.24	1.82	1.43	1.04	0.59	0.21	–	–
d	%	–	42.5	42.6	42.8	43.5	43.6	44.2	45.0	45.3	46.0	47.0	47.5	48.5
	I	8.00	7.75	7.70	7.59	7.40	7.37	7.18	6.97	6.88	6.75	6.50	6.40	6.20
e	%	–	79.0	80.0	80.5	81.9	83.0	84.1	85.5	87.0	88.5	90.0	91.8	93.5
	I	2.00	1.92	1.82	1.76	1.62	1.51	1.40	1.26	1.13	0.97	0.85	0.70	0.55
f	%	–	63.0	64.5	65.7	67.5	69.2	70.6	72.5	75.0	77.0	79.0	81.3	84.3
	I	4.00	3.82	3.63	3.46	3.23	3.02	2.84	2.62	2.34	2.15	1.92	1.68	1.38
g	%	–	63.4	65.8	68.3	71.5	74.5	78.0	81.9	85.0	88.3	93.3	98.0	–
	I	4.00	3.75	3.48	3.13	2.75	2.40	2.02	1.63	1.32	1.01	0.58	0.17	–

Initial rates of reaction are given by the slopes (with sign reversed) at time zero of the curves shown on page 82.

a $\dfrac{(4.0 - 2.8) \times 10^{-3}\ \text{mol dm}^{-3}}{6.0\ \text{min}} = 2.0 \times 10^{-4}\ \text{mol dm}^{-3}\ \text{min}^{-1}$

b $\dfrac{(4.0 - 1.8) \times 10^{-3}\ \text{mol dm}^{-3}}{6.0\ \text{min}} = 3.7 \times 10^{-4}\ \text{mol dm}^{-3}\ \text{min}^{-1} \simeq 2 \times \text{a}$

c $\dfrac{(4.0 - 0.2) \times 10^{-3}\ \text{mol dm}^{-3}}{6.0\ \text{min}} = 6.3 \times 10^{-4}\ \text{mol dm}^{-3}\ \text{min}^{-1} \simeq 3 \times \text{a}$

d $\dfrac{(8.0 - 6.7) \times 10^{-3}\ \text{mol dm}^{-3}}{6.0\ \text{min}} = 2.2 \times 10^{-4}\ \text{mol dm}^{-3}\ \text{min}^{-1} \simeq 1 \times \text{a}$

e $\dfrac{(2.0 - 0.9) \times 10^{-3}\ \text{mol dm}^{-3}}{6.0\ \text{min}} = 1.8 \times 10^{-4}\ \text{mol dm}^{-3}\ \text{min}^{-1} \simeq 1 \times \text{a}$

f $\dfrac{(4.0 - 1.9) \times 10^{-3}\ \text{mol dm}^{-3}}{6.0\ \text{min}} = 3.5 \times 10^{-4}\ \text{mol dm}^{-3}\ \text{min}^{-1} \simeq 2 \times \text{a}$

g $\dfrac{(4.0 - 0.4) \times 10^{-3}\ \text{mol dm}^{-3}}{6.0\ \text{min}} = 6.0 \times 10^{-4}\ \text{mol dm}^{-3}\ \text{min}^{-1} = 3 \times \text{a}$

Orders of reaction

a, b and c – Initial values of $[I_2(aq)]$ and $[H^+(aq)]$ are constant. Initial values of $[CH_3COCH_3(aq)]$ and initial rates are both in ratio 1:2:3. Therefore, reaction is first order with respect to CH_3COCH_3.

a, d and e – Initial values of $[CH_3COCH_3(aq)]$ and $[H^+(aq)]$ are constant. Initial values of $[I_2(aq)]$ vary but initial rate is constant. Therefore, reaction is zero order with respect to I_2.

a, f and g – Initial values of $[CH_3COCH_3(aq)]$ and $[I_2(aq)]$ are constant. Initial values of $[H^+(aq)]$ and initial rates are both in ratio 1:2:3. Therefore, reaction is first order with respect to H^+.

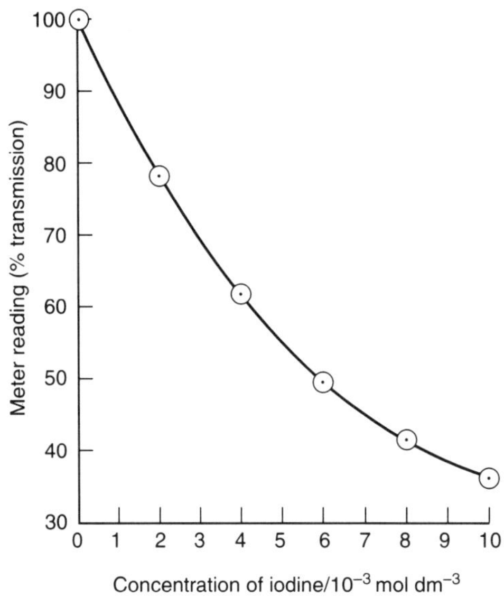

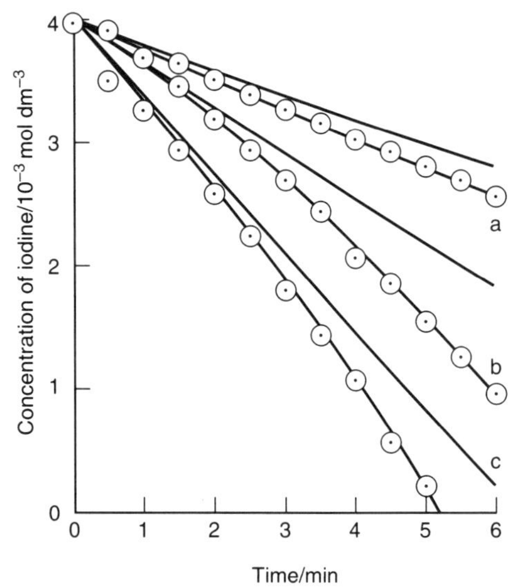

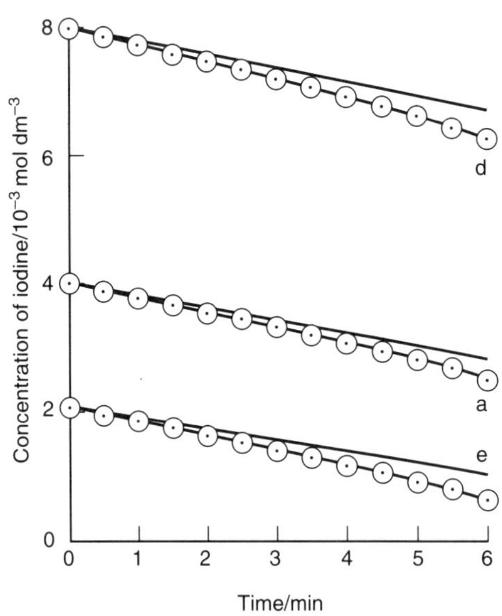

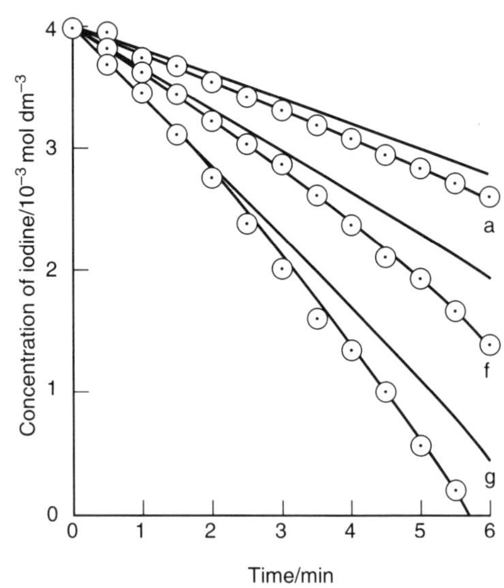

Answers to questions

1. The graphs show that the rate of reaction is directly proportional to the initial concentrations of both propanone and hydrogen ions, but is independent of the initial concentration of iodine. Thus, the reaction is first order with respect to propanone, first order with respect to hydrogen ions and zero order with respect to iodine. The rate equation is:

$$\text{rate} = k[CH_3COCH_3(aq)][H^+(aq)]$$

2. The overall order of reaction is 2.
3. Hydrogen ions appear in the rate equation but are not reactants in the stoichiometric equation. On the other hand, iodine is a reactant but does not appear in the rate equation.
   The rate of reaction must be determined by a step in the mechanism which involves propanone and hydrogen ions but not iodine.
4. Rate constants for this reaction at different temperatures have been quoted as follows:

Temperature/°C	15	20	25	30	35	40
Rate constant/10^{-3} mol^{-1} dm^{3} min^{-1}	0.49	0.90	1.63	2.90	5.07	8.70

An average value at about 23 °C is obtained from our specimen results by dividing initial rates by initial concentrations as follows:

a $\frac{2.0 \times 10^{-4}\ \text{mol dm}^{-3}\ \text{min}^{-1}}{(0.40 \times 0.40)\ \text{mol}^{2}\ \text{dm}^{-6}} = 1.25 \times 10^{-3}\ \text{mol}^{-1}\ \text{dm}^{3}\ \text{min}^{-1}$

b $\frac{3.7 \times 10^{-4}\ \text{mol dm}^{-3}\ \text{min}^{-1}}{(0.80 \times 0.40)\ \text{mol}^{2}\ \text{dm}^{-6}} = 1.16 \times 10^{-3}\ \text{mol}^{-1}\ \text{dm}^{3}\ \text{min}^{-1}$

c $\frac{6.3 \times 10^{-4}\ \text{mol dm}^{-3}\ \text{min}^{-1}}{(1.20 \times 0.40)\ \text{mol}^{2}\ \text{dm}^{-6}} = 1.31 \times 10^{-3}\ \text{mol}^{-1}\ \text{dm}^{3}\ \text{min}^{-1}$

d $\frac{2.2 \times 10^{-4}\ \text{mol dm}^{-3}\ \text{min}^{-1}}{(0.40 \times 0.40)\ \text{mol}^{2}\ \text{dm}^{-6}} = 1.38 \times 10^{-3}\ \text{mol}^{-1}\ \text{dm}^{3}\ \text{min}^{-1}$

e $\frac{1.8 \times 10^{-4}\ \text{mol dm}^{-3}\ \text{min}^{-1}}{(0.40 \times 0.40)\ \text{mol}^{2}\ \text{dm}^{-6}} = 1.13 \times 10^{-3}\ \text{mol}^{-1}\ \text{dm}^{3}\ \text{min}^{-1}$

f $\frac{3.5 \times 10^{-4}\ \text{mol dm}^{-3}\ \text{min}^{-1}}{(0.40 \times 0.80)\ \text{mol}^{2}\ \text{dm}^{-6}} = 1.09 \times 10^{-3}\ \text{mol}^{-1}\ \text{dm}^{3}\ \text{min}^{-1}$

g $\frac{6.0 \times 10^{-4}\ \text{mol dm}^{-3}\ \text{min}^{-1}}{(0.40 \times 1.20)\ \text{mol}^{2}\ \text{dm}^{-6}} = 1.25 \times 10^{-3}\ \text{mol}^{-1}\ \text{dm}^{3}\ \text{min}^{-1}$

Mean value 1.22×10^{-3} mol^{-1} dm^{3} min^{-1}

Students' values for the rate constant may differ considerably from those we have quoted, for a number of reasons (see answer to question 5, below). Nevertheless, they should be able to deduce the rate equation without too much difficulty.

5. The main sources of error may include the following.
   a Temperature variation during the experiment.
   b Variation in light output from the bulb in the colorimeter.
   c Sample tubes transmitting light differently. This may be due to:
   i) different tubes having different optical characteristics,
   ii) the same tube being in different positions relative to the bulb, and behaving as a lens,
   iii) drops of liquid or patches of dirt on the outside of the tube.
   d Solutions being prepared inaccurately.

EXPERIMENT 9.6 Determining the activation energy of a reaction

This experiment is straightforward and should give good results if the procedure is followed carefully. Make sure your students understand that they need not waste time adjusting the temperature precisely to the values we have stated, but that they must record the actual temperatures carefully. Note, however, that at temperatures higher than 60 °C, the starch–iodine complex is broken down (reversibly) and the blue colour does not appear! This would be a suitable experiment for assessed practical course work. A suggested mark scheme and specimen results are given below.

You may wish to point out to your students that plotting ln t instead of ln($1/t$) is simpler and only changes the sign of the slope.

Suggested mark scheme

Total marks for manipulation = 7, as follows:

Sets up water bath appropriately	(1)
Correct use of burettes	(1)
Correct positioning of test-tubes in water bath	(1)
Mixes solutions in a way that minimises cooling	(1)
Starts clock with minimum delay after mixing	(1)
Stops clock immediately blue colour appears	(1)
Maintains temperatures at values close to those stated	(1)

Total mark	7	6	5–4	3	2	1–0
Scaled mark	5	4	3	2	1	0

Total marks for interpretation = 12, as follows:

Recognises correct equation to calculate activation energy	(1)
Converts temperatures into absolute temperatures correctly	(1)
Converts absolute temperatures into reciprocal values correctly	(1)
Converts times to reciprocal values correctly	(1)
Converts reciprocal times to $\log_e$ values correctly	(1)
Plots ln($1/t$) (vertical axis) against $1/T$ (horizontal axis)	(1)
Plots points accurately	(1)
Labels axes correctly	(1)
Obtains best straight line	(1)
Recognises that gradient of graph is equal to $-E/R$	(1)
Calculates activation energy giving correct units	(2)

Total mark	12–11	10–9	8–7	6–5	4–3	2–0
Scaled mark	5	4	3	2	1	0

Specimen results

Results Table 9.8

Temperature/°C	30	36	39	45	51
Temperature, *T*/K	303	309	312	318	324
Time *t*/s	204	138	115	75	55
$\ln\frac{1}{t}$	−5.32	−4.93	−4.75	−4.32	−4.01
$\frac{1}{T}/10^{-3}\,K^{-1}$ (or $10^3\,K/T$)	3.30	3.24	3.21	3.14	3.09

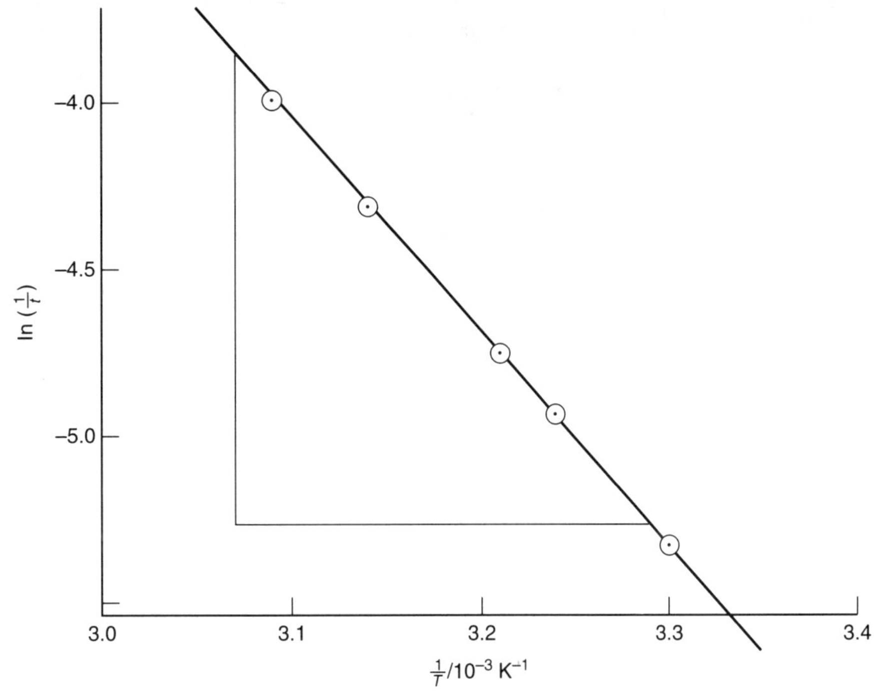

$$\text{Slope of graph} = \frac{\Delta y}{\Delta x} = \frac{-5.26 - (-3.86)}{(3.290 - 3.070) \times 10^{-3}\ \text{K}^{-1}} = -6.36 \times 10^{3}\ \text{K}$$

$$\text{But slope} = \frac{-E_a}{R}$$

$$\therefore E_a = -\text{slope} \times R$$
$$= 6.36 \times 10^{3}\ \text{K} \times 8.31\ \text{J K}^{-1}\ \text{mol}^{-1}$$
$$= \mathbf{52.9\ kJ\ mol^{-1}}$$

EXPERIMENT 9.7 Determining the activation energy of a catalysed reaction

You can use this experiment for the assessment of planning abilities, if you wish. A suggested mark scheme and specimen results are given below. Students would be expected to base their plan on the procedure for Experiment 9.6 but work at lower temperatures. You may wish to suggest using lower concentrations, however, because those specified for Experiment 9.6 give rather short reaction times – see our specimen results.

Suggested mark scheme

Total marks for planning = 8, as follows:

Prepares suitable list of apparatus required	(1)
Prepares suitable list of chemicals required	(1)
Chooses suitable quantities of solutions	(1)
Chooses suitable concentrations of solutions, e.g. 0.01 M $K_2S_2O_8$	(1)
Chooses suitable range of temperatures, e.g. 15–40 °C	(1)
Chooses suitable procedure, e.g. as in Experiment 9.6	(1)
Writes clear instructions	(1)
Prepares suitable results table	(1)

Total mark	8	7–6	5–4	3	2–1	0
Scaled mark	5	4	3	2	1	0

Specimen results

Temperature/°C	15	19.5	26	35	42
Temperature, *T*/K	288	292.5	299	308	315
Time, *t*/s	10.0	7.0	5.0	3.5	2.5
ln(1/*t*)	−2.30	−1.95	−1.61	−1.25	−0.92
$\frac{1}{T}/10^{-3}\ K^{-1}$	3.472	3.419	3.344	3.247	3.175

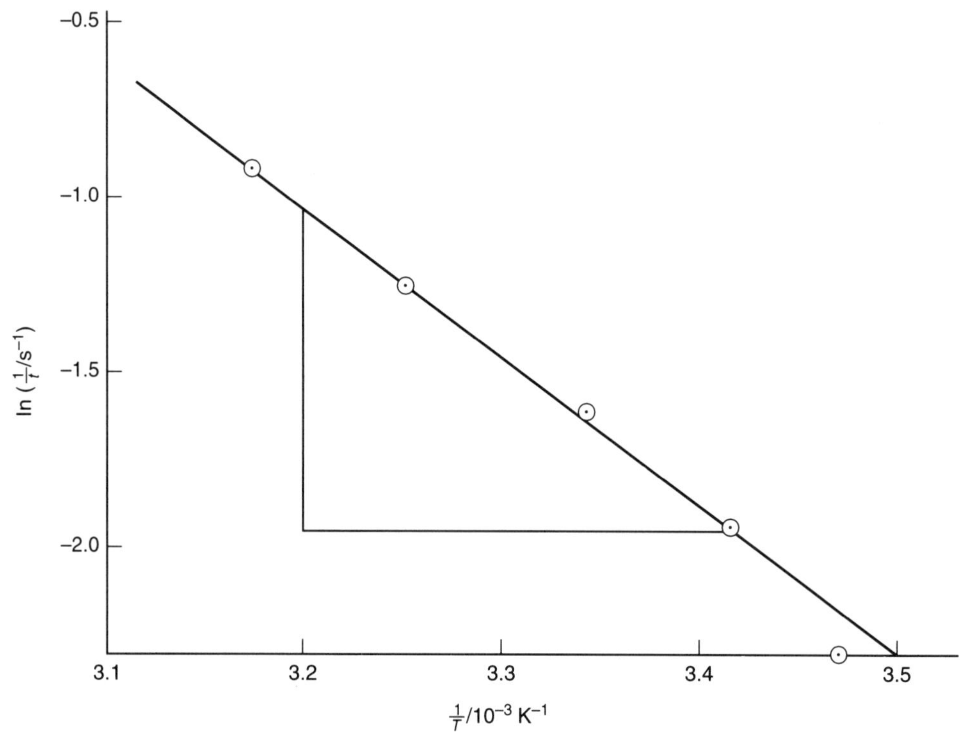

$$\text{Slope} = \frac{-1.96 - (-1.04)}{(3.420 - 3.200) \times 10^{-3}\ K^{-1}} = \frac{-0.92}{0.220 \times 10^{-3}\ K^{-1}} = -4180\ K$$

But the slope $= -\dfrac{E_a}{R}$

$\therefore E_a = -$ slope $\times R$

$= 4180\ K \times 8.31\ J\ K\ mol^{-1}$ = $\mathbf{35\ kJ\ mol^{-1}}$

Despite the errors involved in measuring the short reaction times, this result is close to the quoted value of 28 ± 2 kJ mol^{-1} (*School Science Review*, 1975, **57**,(198), page 108).

EXPERIMENT 9.8 A bromine 'clock' reaction

This experiment is more straightforward than it appears and should give good results provided students prepare their reaction mixtures carefully and mix them thoroughly. Some care is needed in the preparation of solutions by the technician – in particular, the phenol solution is so dilute that its concentration cannot be relied upon unless it is made up just before the experiment by dilution of a more concentrated stock solution. Students may pool their results to save time – temperature variation between sets of results does not matter as long as it is constant within each set.

Specimen results

Results Table 9.9

Volume of $Br^-(aq)/cm^3$	10.0	8.0	6.0	5.0	4.0	3.0
t/s	22.5	26.0	34.0	37.5	48.0	65.0
$\frac{1}{t}/10^{-2}\ s^{-1}$	4.44	3.85	2.94	2.67	2.08	1.54
Temperature/°C	21.0	21.5	21.5	21.5	22.0	22.0

Average temperature of solutions = 21.5 °C.

Results Table 9.10

Volume of $BrO_3^-(aq)/cm^3$	10.0	8.0	6.0	5.0	4.0	3.0
t/s	18.5	24.0	31.0	35.0	45.0	69.0
$\frac{1}{t}/10^{-2}\ s^{-1}$	5.41	4.17	3.23	2.86	2.22	1.45
Temperature/°C	21.5	22.0	21.5	21.5	22.0	21.5

Average temperature of solutions = 21.5 °C.

Results Table 9.11

Volume of acid/cm^3	10.0	8.0	6.0	5.0	4.0	3.0
t/s	13.0	19.0	30.0	39.0	60.0	97.0
$\frac{1}{t}/10^{-2}\ s^{-1}$	7.69	5.26	3.33	2.56	1.67	1.03
Temperature/°C	21.5	21.5	21.5	22.0	21.5	21.5

Average temperature of solutions = 21.5 °C.

Graphs 1 and 2 are straight lines (see below). Therefore the reaction is first order with respect to $Br^-(aq)$ and to $BrO_3^-(aq)$. Graph 3 is **not** a straight line. (Note that the line must pass through the origin since, when $V = 0$, $t = \infty$ and $1/t = 0$.)

Plotting $\log \frac{1}{t}$ against log (volume of acid/cm^3):

$\log\left(\frac{1}{t}/10^{-2}\ s^{-1}\right)$	0.886	0.721	0.522	0.408	0.223	0.013
$\log(V/cm^3)$	1.000	0.903	0.778	0.699	0.602	0.477

Graph 1

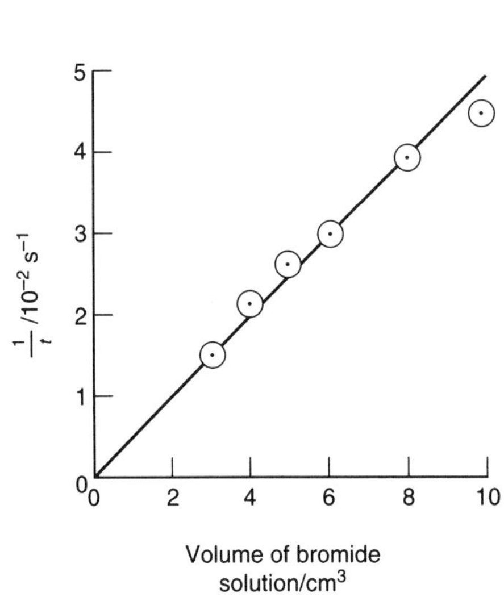

Graph 2

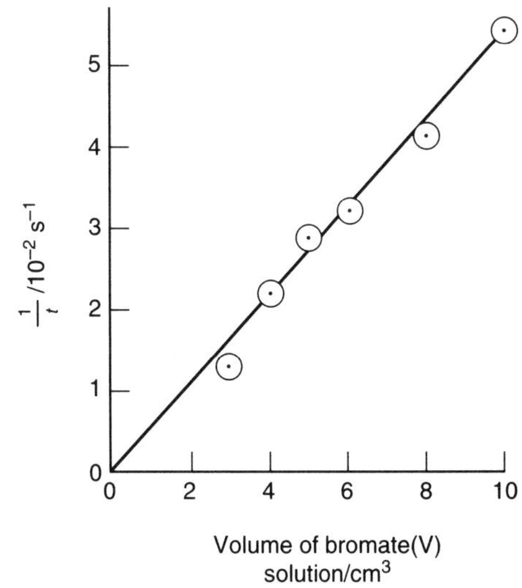

Graph 3

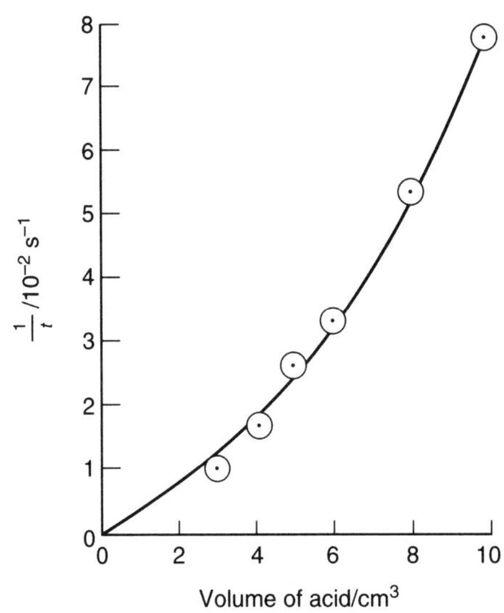

Graph 4

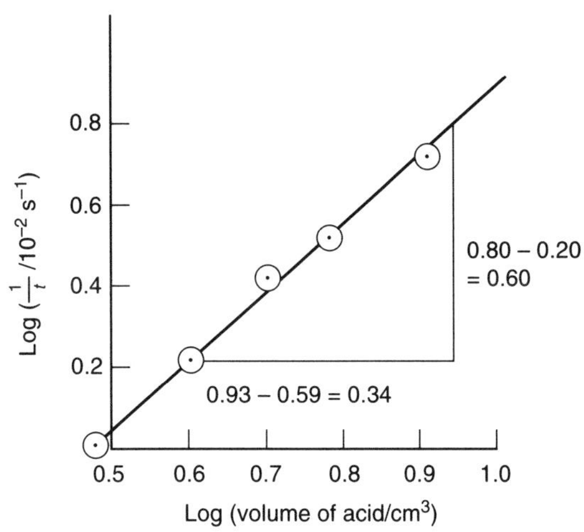

Slope of graph 4 = $\dfrac{0.60}{0.34} = 1.8$

This suggests that the reaction is second order with respect to $H^+(aq)$. (However, it is possible that there could be two mechanisms operating at once, which could give a fractional order of 1.8.)

Answers to questions

1. Rate = $k[Br^-(aq)][BrO_3^-(aq)][H^+(aq)]^2$
2. The reaction producing bromine:

$$5Br^-(aq) + BrO_3^-(aq) + 6H^+(aq) \rightarrow 3Br_2(aq) + 3H_2O(l)$$

is slow compared with the other two. This allows the bromine to be removed from solution as soon as it is formed, by the reaction with phenol:

$$3Br_2(aq) + C_6H_5OH(aq) \rightarrow C_6H_2Br_3OH(aq) + 3H^+(aq) + 3Br^-(aq)$$

Similarly, the indicator reaction is fast so that as soon as the phenol has been used up, free bromine will bleach its colour.

3. The phenol solution needs to be dilute in order to restrict the main reaction to its early stages, when the concentration–time curve is almost linear. Within this region the average rate of reaction (which is what you measure in a clock reaction) approximates to its initial rate.

Note on the mechanism

Students may be surprised to find a reaction which is fourth order overall; this seems to imply that four molecules collide in the rate-determining step – a most unlikely occurrence!

A possible mechanism is:

$H^+(aq) + Br^-(aq) \rightleftharpoons HBr(aq)$	fast
$H^+(aq) + BrO_3^-(aq) \rightleftharpoons HBrO_3(aq)$	fast
$HBr(aq) + HBrO_3(aq) \rightarrow$ products	slow

Then rate = $k[HBr(aq)][HBrO_3(aq)]$

But $[HBr(aq)] \propto [H^+(aq)][Br^-(aq)]$
and $[HBrO_3(aq)] \propto [H^+(aq)][BrO_3^-(aq)]$ (from consideration of equilibrium constants)

$\therefore$ rate = $k^1[Br^-(aq)][BrO_3^-(aq)][H^+(aq)]^2$

EXPERIMENT 10.1 Chemical properties of carboxylic acids

In this experiment the students perform some simple test-tube reactions with glacial ethanoic acid. Aqueous ethanoic acid must **not** be used because the reactions of water with sodium and phosphorus pentachloride not only are violent, but also mask the reactions of the —OH group in the acid, which are more moderate. You should closely supervise the use of sodium (or demonstrate if you wish) and not allow students to cut their own pieces from a large lump.

The triiodomethane (iodoform) reaction may fail to give a yellow precipitate if an iodine solution less concentrated than 10% is used.

From their observations, the students should conclude that the presence of an —OH group on the carbonyl carbon atom modifies the properties of both groups; carboxylic acids show virtually none of the properties of aldehydes and ketones and they are much more acidic than alcohols.

Specimen results

Results Table 10.1
Reactions of ethanoic acid

	Reaction	Observations
A	pH of aqueous solution	Orange–red – pH 3–4
B	Reaction with sodium hydrogen-carbonate solution	Gas evolved which turns lime-water milky – CO_2
C	Reaction with sodium	Gas evolved which popped with a lighted splint – H_2
D	Reaction with phosphorus pentachloride	Steamy gas evolved which gave white fumes with ammonia – HCl
E	Reaction with 2,4-dinitro-phenylhydrazine	No change
F	Triiodomethane reaction	No change
G	Action of iron(III) chloride	Red colour at first – brown ppt. on boiling

Answers to questions

1. Ethanoic acid resembles hydroxy compounds more closely than carbonyl compounds. It reacts with sodium and phosphorus pentachloride, typical reactions of alcohols, but does not react with 2,4-dinitrophenylhydrazine.
2. Ethanoic acid releases carbon dioxide from sodium hydrogencarbonate, but phenol does not. Also, the pH of aqueous ethanoic acid is lower than that of aqueous phenol of similar concentration.
3. The presence of the large, non-polar benzene ring outweighs the influence of the polar carboxyl group, which participates in hydrogen-bonding with water.

EXPERIMENT 10.2 Chemical properties of ethanoyl chloride

In this experiment the students add ethanoyl chloride to water, ethanol, ammonia and phenylamine. Since some of these reactions are violent, especially those with ammonia and phenylamine, make sure your students follow the instructions carefully. They should perform the reactions in a fume cupboard with the safety glass pulled down as far as is practically possible. In view of the possible hazards in performing this experiment you may decide to carry it out yourself as a demonstration. It should be pointed out to the students that the concentrated ammonia solution should not be used for test A.3 – only for test C.

Specimen results

Results Table 10.2

Reaction		Observations
A	**Reaction with water**	Moderate reaction giving steamy fumes
	Product tested with ammonia	White fumes, suggesting HCl is produced
	Product tested with iron(III) chloride	Red coloration, suggesting ethanoic acid is formed
B	**Reaction with ethanol**	Vigorous reaction with crackling and spitting, giving steamy fumes
	Product tested with ammonia	White fumes, suggesting HCl is produced
	Smell of product	Sweetish smell, suggesting an ester is produced
C	**Reaction with ammonia**	Violent reaction, with crackling and spitting, giving a lot of white fumes (NH_4Cl)
D	**Reaction with phenylamine**	Violent reaction with crackling and spitting, giving a lot of white fumes. Some white solid around lower walls of beaker

Answers to questions

1. **a** $CH_3COCl + H_2O \rightarrow CH_3CO_2H + HCl$
   ethanoic acid

   b $CH_3COCl + C_2H_5OH \rightarrow CH_3CO_2C_2H_5 + HCl$
   ethyl ethanoate

   c $CH_3COCl + 2NH_3 \rightarrow CH_3CONH_2 + NH_4Cl$
   ethanamide

   d $CH_3COCl + C_6H_5NH_2 \rightarrow CH_3CONHC_6H_5 + HCl$ (HCl reacts with excess amine)
   N-phenylethanamide

2. The order of reactivity of the nucleophiles corresponds with the order of basic strength.

$$NH_3 > C_6H_5NH_2 > C_2H_5OH > H_2O$$

The more basic the nucleophile, the more readily available is the lone pair of electrons to make a bond with ethanoyl chloride.

EXPERIMENT 10.3 Preparation of aspirin

This experiment may be used for practical assessment of the skills in manipulation. This experiment also gives students a second chance to perform the techniques of recrystallisation and melting point determination, which they meet in Experiment 8.5. During the crystallisation procedure, your students may find that they need to add quite a lot of water to dissolve the aspirin. In this case, they may find it more convenient to use a small beaker rather than a boiling-tube. If time is short and your particular syllabus does not specify that your students should know the details of the laboratory preparation of an ester, they could omit this experiment. An alternative experiment (Experiment 10.9) is to prepare the ester phenyl benzoate from phenol and benzoyl chloride.

Suggested mark scheme

Total marks for manipulation = 15, as follows:

Handles apparatus

- with dexterity (5)
- intermediate (4)
- with some skill (3)
- intermediate (2)
- awkwardly (1)

Yield/g	Mark
1.8	5
1.5–1.8	4
1.3–1.5	3
1.0–1.3	2
0.75–1.0	1

Melting point = 135 °C	Mark
within 2°	5
within 4°	4
within 6°	3
within 8°	2
within 10°	1

Total mark	15–13	12–11	10–9	8–6	5–4	3–0
Scaled mark	5	4	3	2	1	0

EXPERIMENT 10.4 Identification of two organic compounds

This is intended either for students taking an A-level practical examination in which one of the questions involves the identification of an unknown compound or for students being internally assessed on their practical work.

F = 4-oxopentanoic acid (levulinic acid), CH_3COCH_2COOH*

G = benzaldehyde C_6H_5O

* This may not be available from your usual supplier. Try Sigma-Aldrich Chemical Co Ltd, The Old Brick Yard, New Road, Gillingham, Dorset, SP8 4XT. Tel. FREEFONE (0800) 373731. Stock number L0626. Price (1997) 50 g £5.30, 250 g £10.90 (£6.50 carriage charge).

Note: The procedure for the triidomethane reaction in test **d** is not the one we have followed in our experiments. The test may fail if the sodium chlorate(I) (hypochlorite) solution is not fresh; we suggest you check.

Suggested mark scheme

Total marks for observing and recording = 9.
Record marks from observations in Results Table 10.4.

Total mark	9–8	7–6	5–4	3	2	1–0
Scaled mark	5	4	3	2	1	0

Total marks for interpreting = 24.
Record marks from inferences in Results Table 10.4 and from the answers to questions.

Total mark	24–22	21–18	17–14	13–10	9–6	5–0
Scaled mark	5	4	3	2	1	0

Specimen results

Results Table 10.4

Experiment	Observations	Inferences
a To 2 cm^3 of the solution of F add 2,4-dinitrophenylhydrazine reagent	Yellow precipitate formed (1)	A condensation reaction occurs (1). This shows F is an aldehyde or ketone (1)
b Prepare a sample of Tollens reagent as follows: to 5 cm^3 of aqueous silver nitrate in a test-tube add one to two drops of aqueous sodium hydroxide. Then add dilute aqueous ammonia until only a trace of precipitate remains. Now add five drops of the solution of F and place the tube in hot water. (Pour the contents of the tube down the sink on completion of the test)	No change observed (1)	No reduction of silver(I) to silver metal (1). Therefore, F is not an aldehyde (1)
c Add a little of the solution of F to some sodium hydrogencarbonate	Effervescence (1). Gas turned lime-water cloudy (1)	Carbon dioxide released by acid (1). F probably contains a carboxyl group, CO_2H (1)
d To 1 cm^3 of the solution of F add 3 cm^3 of aqueous potassium iodide and then 10 cm^3 of sodium chlorate(I) (sodium hypochlorite) solution	Yellow precipitate formed (1)	Triiodomethane produced (1). This shows the presence of structural group $CH_3\underset{\mid}{C}=O$ or $CH_3\underset{\mid}{C}HOH$ (1)
e Place **one** drop of G on an inverted crucible lid and ignite G from above	G burned with a sooty (1) luminous flame (1)	G is probably an aromatic compound (1)
f To 1–2 cm^3 of G add 2,4-dinitrophenylhydrazine reagent	An orange–yellow precipitate was formed (1)	A condensation reaction occurs (1). This shows G is an aldehyde or ketone (1)
g Prepare a sample of Tollens reagent as in **b**. Add five drops of G, shake the mixture and place the tube in hot water. (Pour the contents of the tube down the sink on completion of the test)	A silver mirror was formed slowly on the inner surface of the tube (1)	Formation of silver metal from silver(I) shows G is a reducing agent (1), probably an aldehyde (1)

Answers to questions

1. Since F is acidic (test **c**) it probably contains a carboxyl group, —CO_2H. (2)
   Since F forms a 2,4-dinitrophenylhydrazone (test **a**) it must contain a carbonyl group, $>C=O$ in addition to the —CO_2H group. (2)
   Since F undergoes the triiodomethane reaction (test **d**) it must contain one of the structural groupings $CH_3C=O$ or CH_3CHOH. (2)
2. Test **e** suggests that G is aromatic. (1)
   The formation of a 2,4-dinitrophenylhydrazone in test **f** shows the presence of a carbonyl group, $>C=O$. (2)
   The formation of a silver mirror in test **g** suggests that the carbonyl group is in an aldehyde, —CHO. (2)

EXPERIMENT 10.5 The glycine/copper(II) complex

This is a very simple experiment and should take only about 10 minutes. We have assumed that students have dealt with complex ions already, but if this is not the case, they will need some help in answering the questions.

Answers to questions

1. The solution would be slightly alkaline. Since CO_3^{2-} is the conjugate base of the weak acid HCO_3^- which is, in turn, the conjugate base of the weak acid H_2CO_3, the following equilibria would be set up:

$$CO_3^{2-}(aq) + H^+(aq) \rightleftharpoons HCO_3^-(aq)$$
$$HCO_3^-(aq) + H^+(aq) \rightleftharpoons H_2CO_3(aq)$$

   These would decrease the concentration of hydrogen ions in solution, making it slightly alkaline.
2. In alkaline solution, the zwitterion form of glycine would tend to lose a proton and become the basic form:

$$CH_3NH_3^+CO_2^-(aq) + OH^-(aq) \rightleftharpoons CH_2NH_2CO_2^-(aq) + H_2O(l)$$

3. The basic form of glycine can make two dative bonds, using the lone pair of electrons on the nitrogen and the negative charge on the carboxyl oxygen.

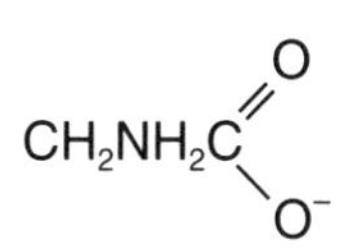

   Glycine is therefore a bidentate ligand.
4.

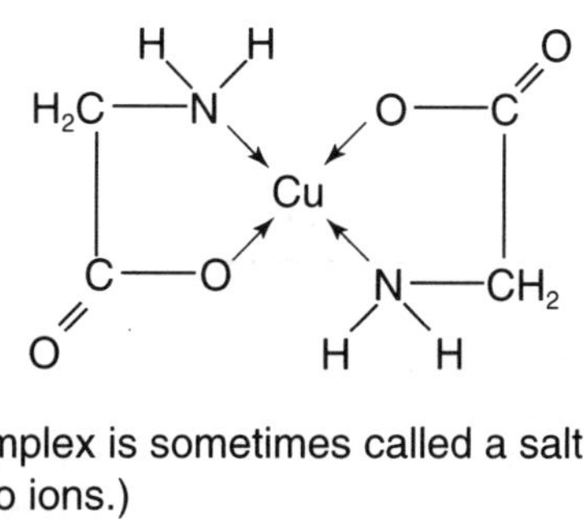

   (Note that this complex is sometimes called a salt because it arises from the combination of two ions.)
5. This structure would be essentially square planar and would have no net charge.

EXPERIMENT 10.6 Biuret test for proteins

This experiment also is very short and straightforward but, as before, some knowledge of complex formation is assumed. Make sure that students appreciate the difference (in colour and in structure) between the complexes formed in this experiment and the one in Experiment 10.5. The biuret test is given by all proteins and all other compounds containing two or more peptide links.

Answers to questions

1. The intensity of the purple (or pink) colour is related to the concentration of the protein. A colorimeter or spectrophotometer could be calibrated to enable concentration to be determined from colour intensity.
2.

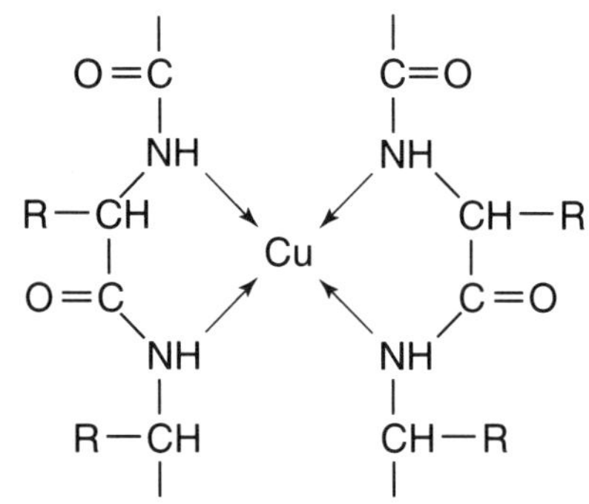

3. No, because the biuret complex is formed from nitrogen in peptide groups. However, care is needed because some amino acids form bluish complexes with copper(II) ions (see Experiment 10.5) and the colours can be confused.

EXPERIMENT 10.7 Paper chromatography

We have included this simple separation of amino acids for those students who have had little or no experience of chromatography in their pre-A-level course. It uses simple apparatus, small quantities of solvent and can be completed in one hour.

Make sure that students are aware of the hazards of using ninhydrin as a locating agent.

Specimen results

Results Table 10.5

Amino acid		Distances travelled/cm		
		By solvent	By amino acid	R_f value
Aspartic acid	– alone	7.1	1.1	0.15
	– in mixture	6.8	0.9	0.13
Leucine	– alone	7.0	5.5	0.79
	– in mixture	6.8	5.1	0.75
Lysine	– alone	7.1	2.6	0.37
	– in mixture	6.8	2.4	0.35

Answers to questions

1. R_f values vary slightly with local conditions such as temperature, the purity of the solvent, the nature of the chromatography paper, air pressure and humidity.
2. The greater the solubility of a substance in a particular solvent, the less readily it is retained by the stationary phase and the faster it appears to move. This gives a higher R_f value.
3. Amino acids are readily transferred from moist and/or dirty fingers on to the paper and show up as confusing brown marks when sprayed with ninhydrin solution.

EXPERIMENT 10.8 Identification of three organic compounds

This experiment can be used for students being internally assessed on observing and recording and interpretation skills.

X – Benzoic acid
Y – Propan-2-ol
Z – Butanone

Part A requires chemical testing. Part B requires the students to interpret IR and mass spectra and should only be attempted after they have completed and handed in part A.

Only part A is intended for assessment purposes. For part B the students will need access to data on melting points and boiling points of organic compounds. They will also need copies of pages 247 to 249, which give data on % elemental analysis, mass spectra and IR spectra for the three unknown compounds. You will need to warn the students that compounds Y and Z are highly flammable and the vapours are harmful. The boiling point determinations **must be done in the fume cupboard**.

Suggested mark scheme

Total marks for observing and recording = 18.
Record marks from observations in Results Table 10.6.

Total mark	18–16	15–13	12–11	10–8	7–5	4–0
Scaled mark	5	4	3	2	1	0

Total marks for interpreting = 20.
Record marks from inferences in Results Tables 10.6 and 10.7.

Total mark	20–17	16–14	13–11	10–7	6–4	3–0
Scaled mark	5	4	3	2	1	0

Specimen results Part A Chemical Testing

Results Table 10.6

Test		Observations	Inferences
1. **a**	Burn a small amount of X in a combustion spoon	Sooty flame (1)	High C:H ratio (1) probably contains a benzene ring (1)
b	Shake a small amount of X in a test-tube with 5 cm^3 of water. First try cold water then hot. (Use a hot water-bath)	(i) Slightly soluble in cold water (1) (ii) Fully soluble in hot water (1)	Polar group(s) present (1)
c	To a small amount of X in a test-tube add 5 cm^3 of 0.5 mol dm^{-3} sodium carbonate solution	Bubbles of colourless gas (CO_2) (1)	Carboxyl group —CO_2H present (1)
d	Determine the melting point of X	Melting point of X = 120–122 °C (2) (only one mark if 118–119 °C or 123–124 °C)	
2. **a**	Burn a small volume of Y in a combustion spoon	Smokeless flame (1)	Low C:H probably no benzene ring present (1)
b	Shake a small amount (1–2 cm^3) of Y with 5 cm^3 of water in a test-tube. First try cold water then hot. (Use a hot water-bath)	(i) Soluble in cold water (1) (ii) Very soluble in hot water (1)	Polar groups' present (1)
c	To a few drops of Y in a test-tube, add about 5 cm^3 of a mixture of aqueous potassium dichromate(VI) and 2M H_2SO_4 and warm in a hot water-bath	Mixture changes from orange to green (1)	Primary or secondary alcohol or an aldehyde (1)
d	Determine the boiling point of Y in a fume cupboard	Boiling point of Y 82–84 °C (2) (only one mark if 80 °C, 81 °C or 85 °C, 86 °C	
3. **a**	Burn a small volume of Z on a combustion spoon	Smokeless flame (1)	Probably no benzene ring present (1)
b	Shake a small amount (1–2 cm^3) of Z with 5 cm^3 of water in a test-tube. First try cold then hot water	(i) Fairly soluble in cold water (1) (ii) Fully soluble in hot water (1)	Polar group(s) present (1)
c	Add a few drops of Z to 5 cm^3 of a solution of 2,4-dinitrophenyl-hydrazine (Brady's reagent) in a test-tube	Yellow/orange precipate (1)	Carbonyl group (C=O) presence of aldehyde or ketone (1)
d	Determine the boiling point of Z in a fume cupboard	Boiling point of Z 79–81 °C (2) (only one mark if 77 °C, 78 °C or 82 °C, 83 °C	

Results Table 10.7

Compound	Functional group	Reasoning
X	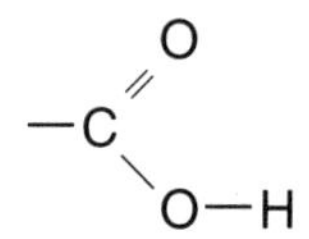 (1) carboxyl group 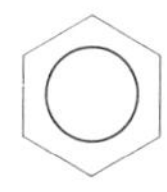 (1) benzene ring	Evolves CO_2 from Na_2CO_3 (1), sooty flame indicates aromatic (benzene ring present) (1)
Y	$\rangle$C=O or —OH (1) (1)	Acidified sodium dichromate, orange→green, either primary or secondary alcohol (1) or aldehyde (1)
Z	$\rangle$C=O (1)	Orange precipitate with 2,4-dinitrophenylhydrazine indicates a carbonyl group of an aldehyde or ketone (1)

Part B

Compound X Empirical formula: $C_7H_6O_2$

Mass spectrum: Molecular ion peak *m/e* 122 gives molecular formula $C_7H_6O_2$.
m/e 77 = $C_6H_5^+$ (suggests aromatic)
m/e 105 = $C_6H_5CO^+$

IR spectrum: The very broad dip at about 3000 cm^{-1} is due to an O—H group.
The dip at about 1700 cm^{-1} is due to a C=O group.
The dip between 700 and 800 cm^{-1} is probably due to C—H stretch in a benzene ring.

The two absorptions in the region 700–800 cm^{-1} are possible sources of confusion. These might have indicated a carbon–halogen bond but are probably due to the benzene ring.

All this suggests the structure of compound X to be benzoic acid, which is confirmed by the melting point of 121 °C.

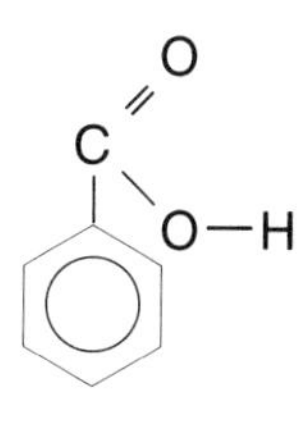

benzoic acid

Compound Y Empirical formula: C_3H_8O

Mass spectrum: Molecular ion peak *m/e* 60 gives molecular formula C_3H_8O.
m/e 27 = $C_2H_3^+$, *m/e* 39 = $C_3H_3^+$, *m/e* 43 = $C_3H_7^+$ or CH_3CO^+, *m/e* 45 = $C_2H_4OH^+$

IR spectrum: The broad dip between 3000 cm^{-1} and 3700 cm^{-1} is due to an O—H group.

Compound Y could be either propan-1-ol or propan-2-ol. The boiling point of 83 °C confirms the structure of Y as:

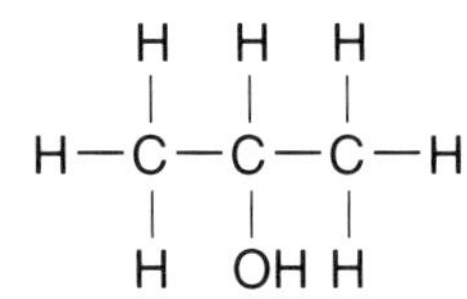

propan-2-ol

Compound Z Empirical formula: C_4H_8O

Mass spectrum: Molecular ion peak at *m/e* = 72 gives molecular formula C_4H_8O.

m/e 43 = CH_3CO^+

m/e 29 = $C_2H_5^+$

IR spectrum: The dip at 1700 cm^{-1} is due to a C=O group, so compound Z is either an aldehyde or a ketone. From the fragments of the mass spectrum above, it is most probably a ketone whose structure is:

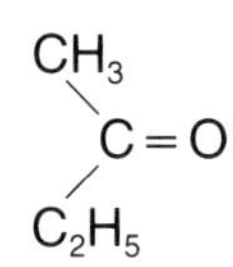

butanone

This is confirmed by the boiling point of 80 °C.

EXPERIMENT 10.9 Preparing phenyl benzoate

This experiment gives students another chance to prepare an ester. This is an alternative to Experiment 10.3 and can be used for practical assessment of the skills in manipulation.

Suggested mark scheme

Total marks for manipulation = 15, as follows:

Handles apparatus

- with dexterity (5)
- intermediate (4)
- with some skill (3)
- intermediate (2)
- awkwardly (1)

Yield/g	Mark
7.0*	5
5.9 – 7.0	4
4.9 – 5.9	3
3.8 – 4.9	2
2.8 – 3.8	1

Melting point = 71 °C*	Mark
within 2°	5
within 4°	4
within 6°	3
within 8°	2
within 10°	1

Total mark	15–13	12–11	10–9	8–6	5–4	3–0
Scaled mark	5	4	3	2	1	0

* You might want to carry out this experiment yourself using the melting point and yield of your sample as a standard for allotting marks.

EXPERIMENT 11.1 The reactions of tin and lead and their aqueous ions

This is a fairly straightforward experiment and works well. Sodium sulphide has been used in place of the more hazardous hydrogen sulphide and gives similar results. Old samples of tin(II) chloride are very difficult to dissolve, even in concentrated hydrochloric acid. If new stock is not available, then an adequate solution can be made by reacting tin with concentrated hydrochloric acid, followed by dilution of the resulting solution.

Suggested mark schemes

Total marks for manipulation = 12, as follows:

Heating **gently** – agitation to avoid spurting, removal from flame to avoid confusion of gas evolution with boiling (2)
Testing for hydrogen – trapping gas where necessary, use of lighted splint, repetition where necessary (3)
Addition of $NaOH/NH_3$ – dropwise addition at first, faster addition to excess (2)
Use of Na_2S – fume cupboard, disposal (2)
General estimation of volumes used (2)
Safety spectacles worn throughout (1)

Total mark	12–10	9–8	7–6	5–4	3–2	1–0
Scaled mark	5	4	3	2	1	0

Total marks for observation and recording = 28, as in Results Tables 11.1 and 11.2 below.

Total mark	28–22	21–18	17–14	13–10	9–5	4–0
Scaled mark	5	4	3	2	1	0

The answers to the questions, as shown below, could be used as the basis for an assessment of interpretation skills but we think it more practicable to use some of the other suggested assessment schemes included in this book (see pages 3 to 4).

Results Table 11.1

Acid	Tin	Lead
Dilute hydrochloric acid	Colourless gas evolved on heating (1), which popped with a lighted splint (1)	Slight reaction (if any) on heating (1)
Concentrated hydrochloric acid	Colourless gas evolved on heating which popped with a lighted splint (1)	Colourless gas evolved on heating. Popped with difficulty (1). White solid produced in test-tube (1)
Concentrated nitric acid	Brown gas evolved on heating (1)	Brown gas evolved on heating (1)

Results Table 11.2

Reagent	Sn^{2+}(aq) (acidified)	Pb^{2+}(aq)
a Sodium hydroxide solution	White precipitate (1) dissolved in excess sodium hydroxide solution (1)	White precipitate (1) dissolved in excess sodium hydroxide solution (1)
b Ammonia solution	White precipitate (1) insoluble in excess ammonia solution (1)	White precipitate (1) insoluble in excess ammonia solution (1)
c Dilute hydrochloric acid	No reaction (1)	White precipitate (1) dissolved on boiling and (1) crystallised on cooling to form white crystals (1)
d Acidified potassium manganate(VII) solution	Decolorised (1)	No reaction (1)
e Potassium chromate(VI) solution	Green–blue solution obtained (1)	Yellow precipitate formed (1)
f Sodium sulphide solution	Brown precipitate formed (1)	Dark brown–black precipitate formed (1)
g Potassium iodide solution	No reaction (1)	Yellow precipitate formed (1)

Answers to questions

1. $Pb(s) + 2HCl(aq) \rightarrow PbCl_2(s) + H_2(g)$ ($PbCl_2(aq)$ if hot)
   $Sn(s) + 2HCl(aq) \rightarrow SnCl_2(aq) + H_2(g)$
2. Yes, metals above hydrogen in the electrochemical series produce hydrogen gas on reaction with acids.
3. **a** Nitrogen dioxide (NO_2) is responsible for the brown colour.
   b Yes, zinc (or copper) reacts in a similar way.
   c Nitric acid is a strong oxidising agent.
4. See Table 11.1 on page 101.
5. If potassium manganate(VII) solution were acidified with hydrochloric acid or sulphuric acid the chloride or sulphate ions would react with the lead ions to form a white precipitate of lead(II) chloride or lead(II) sulphate.

$$Pb^{2+}(aq) + 2Cl^{-} \rightarrow PbCl_2(s)$$
$$Pb^{2+}(aq) + SO_4^{2-} \rightarrow PbSO_4(s)$$

   Ethanoate ions will not form a precipitate with lead ions.
6. Lead(II) ions are more stable because they are not oxidised by any of the reagents used in the experiment. Tin(II) ions are oxidised to the +4 state by acidified solutions of potassium manganate(VII) and potassium dichromate(VI).

Table 11.1

Equations	Comments
$Sn^{2+}(aq) + 2OH^-(aq) \rightarrow Sn(OH)_2(s)$ $Sn(OH)_2(s) + 4OH^-(aq) \rightarrow Sn(OH)_6^{4-}(aq)$	The precipitate dissolves in excess NaOH to form a stannate(II) ion*, $Sn(OH)_6^{4-}$
$Pb^{2+}(aq) + 2OH^-(aq) \rightarrow Pb(OH)_2(s)$ $Pb(OH)_2(s) + 4OH^-(aq) \rightarrow Pb(OH)_6^{4-}(aq)$	The precipitate dissolves in excess NaOH to form a plumbate(II) ion*, $Pb(OH)_6^{4-}$
$2MnO_4^-(aq) + 16H^+(aq) + 5Sn^{2+}(aq) \rightarrow$ $2Mn^{2+}(aq) + 5Sn^{4+}(aq)$*** $+ 8H_2O(l)$	Redox reaction. Sn^{2+} is behaving as a reducing agent
$Pb^{2+}(aq) + 2Cl^-(aq) \rightarrow PbCl_2(s)$	Simple precipitation reaction. Precipitate soluble in hot water
$Cr_2O_7^{2-}(aq)$** $+ 14H^+(aq) + 3Sn^{2+}(aq) \rightarrow$ $2Cr^{3+}(aq) + 3Sn^{4+}(aq)$*** $+ 7H_2O(l)$	Redox reaction. Sn^{2+} is behaving as a reducing agent. Cr^{3+} gives the solution its blue-green colour
$Pb^{2+}(aq) + CrO_4^{2-}(aq) \rightarrow PbCrO_4(s)$	Simple precipitation reaction
$Pb^{2+}(aq) + S^{2-}(aq) \rightarrow PbS$	Simple precipitation reaction
$Sn^{2+}(aq) + S^{2-}(aq) \rightarrow SnS(s)$	Simple precipitation reaction
$Pb^{2+}(aq) + 2I^-(aq) \rightarrow PbI_2(s)$	Simple precipitation reaction. The precipitate is soluble in hot water and recrystallises on cooling as golden yellow platelets

*Various formulae have been proposed for the stannate(II) ion, ranging from SnO_2^{2-} for the anhydrous form to $Sn(OH)_4^{2-}$ and $Sn(OH)_6^{4-}$ for the hydrated forms. $Sn(OH)_6^{4-}$ seems the most probable. Similar variations have been proposed for the plumbate(II) ion, $Pb(OH)_6^{4-}$. **The solution of Sn^{2+} is acidic and chromate(VI), CrO_4^{2-}, changes to dichromate(VI), $Cr_2O_7^{2-}$, when acidified:

$$2CrO_4^{2-}(aq) + 2H^+(aq) \rightleftharpoons Cr_2O_7^{2-}(aq) + H_2O(l)$$

***Sn^{4+} ions are stated in the equations for convenience; you should bear in mind that Sn(IV) compounds are mainly covalent. Aqueous solutions of tin(IV) usually contain complexes

EXPERIMENT 11.2 The preparations and reactions of tin(IV) oxide and lead(IV) oxide

In the preparation of lead(IV) oxide, students should ensure that the product is completely dry, otherwise they will end up with apparent percentage yields in excess of 100%. Since lead(IV) oxide is easily decomposed by heat, we suggest you dry the product in an incubator (borrowed from the biology department) set at about 35 °C, and left for one or two days. Conventional ovens tend to reach much higher temperatures even on the lowest settings. Tin(IV) oxide seems to be particularly 'passive' when prepared in this manner. It is therefore advisable to use samples from the manufacturers for part C of the experiment. It is very difficult to detect whether the oxides dissolve in acids and alkalis by this method and you may need to tell your students that textbook(s) claim that they do.

Suggested mark schemes

Suggested mark schemes for the assessment of manipulative skills are shown below. The two sets could be combined for a small number of students but it may be more practicable to assess some students on part A or part B, or to select only some of the points below for assessment. Some teachers take the view that more weight could be given to the yield since a good yield can only be obtained by good procedures, but this takes no account of safety and could penalise students who performed well for most of the time but lost a lot of product through one error.

Part A

Total marks for manipulation = 10, as follows:

Weighing the tin – appropriate accuracy	(1)
Addition of nitric acid – care to avoid skin contact, fume cupboard, estimates of complete reaction	(3)
Suction filtration – as taught for particular apparatus	(2)
Washing of precipitate	(1)
Heating to dryness – gentle heat at first, stronger when no spitting	(2)
Safety spectacles worn throughout	(1)

Total mark	10–9	8–7	6–5	4	3–2	1–0
Scaled mark	5	4	3	2	1	0

Part B

Total marks for manipulation = 13, as follows:

Weighing the lead – appropriate quantity and accuracy	(2)
Addition of nitric acid – care to avoid skin contact, fume cupboard	(2)
Adjustment of heating to give appropriate rate of reaction	(1)
Addition of water to compensate for evaporation	(1)
Addition of NaOH to give **permanent** precipitate	(1)
Decantation procedure	(1)
Suction filtration – as taught for particular apparatus	(2)
Yield Above 50% (2), above 35% (1)	(2)
Safety spectacles worn throughout	(1)

Total mark	13–11	10–9	8–7	6–5	4–3	2–0
Scaled mark	5	4	3	2	1	0

Specimen results and calculations

Results Table 11.3

Mass of Pb used	5.0 g
Maximum possible mass of PbO_2 obtainable	5.8 g
Mass of PbO_2 obtained	3.9 g
% yield	67%
Appearance of PbO_2	Red–brown solid
Appearance of SnO_2	White/pale yellow solid

Percentage yield of PbO_2 is calculated as follows:

1 mol of Pb produces 1 mol of PbO_2
i.e. 207 g of Pb produces 239 g of PbO_2

$$\therefore \text{ 1 g of Pb produces } \frac{239\text{ g}}{207} \text{ of } PbO_2$$

$$\therefore 5.0 \text{ g of Pb produces } \frac{239 \text{ g}}{207} \times 5.0 = 5.8 \text{ g of } PbO_2 \text{ (theoretical maximum)}$$

$$\% \text{ yield} = \frac{\text{actual mass of product}}{\text{theoretical maximum mass of product}} \times 100$$

$$\therefore \% \text{ yield} = \frac{3.9 \text{ g}}{5.8 \text{ g}} \times 100 = \mathbf{67\%}$$

Results Table 11.4

Test	Tin(IV) oxide	Lead(IV) oxide
a Heat	No change	Red liquid formed on heating and colourless gas evolved which relit a glowing splint (O_2 evolved). Yellow glassy solid remained
b Dilute hydrochloric acid	No reaction	Slow evolution of a gas which bleached damp litmus paper. Characteristic smell of chlorine detectable
c Concentrated hydrochloric acid	No apparent reaction	Vigorous reaction, chlorine evolved, yellow clear liquid remained with a white solid at the bottom of the test-tube
d Concentrated sodium hydroxide	A very small quantity probably dissolved but this was not obvious	A very small quantity dissolved leaving a brown coloured solution. (This may be PbO_2 in suspension)
e Acidified potassium iodide	No reaction	Brown solution formed

Answers to questions

1. $2PbO_2(s) \rightarrow 2PbO(s) + O_2(g)$
2. **a** PbO_2 is oxidising chloride ions to chlorine.

$$PbO_2(s) + 4HCl(aq) \rightarrow PbCl_2(s) + 2H_2O(l) + Cl_2(g)$$

b PbO_2 is oxidising iodide ions to iodine, which dissolves in excess iodide ions.

$$PbO_2(s) + 2I^-(aq) + 2H^+(aq) \rightarrow PbO(s) + I_2(aq) + H_2O(l)$$

3. The yellow liquid contains the complex ion $[PbCl_6]^{2-}$ (hexachloroplumbate(IV) ion).
4. **a** $SnO_2(s) + 2OH^-(aq) + 2H_2O(l) \rightarrow [Sn(OH)_6]^{2-}(aq)$
   b $PbO_2(s) + 2OH^-(aq) + 2H_2O(l) \rightarrow [Pb(OH)_6]^{2-}(aq)$
5. Both oxides (SnO_2 and PbO_2) are amphoteric since they react with acids and alkalis.
6. Oxygen will be evolved when red lead is heated:

$$2Pb_3O_4(s) \rightarrow 6PbO(s) + O_2(g)$$

This reaction can be regarded as the decomposition of the lead(IV) oxide component of red lead, i.e.:

$$2PbO_2(s) \rightarrow 2PbO(s) + O_2(g)$$

7. Tin(II) oxide might be expected to combine with oxygen on heating, to form tin(IV) oxide, because the +4 oxidation state is generally more stable for tin. This happens in practice:

$$2SnO(s) + O_2(g) \rightarrow 2SnO_2(s)$$

A similar oxidation of lead(II) oxide to lead(IV) oxide would **not** be expected because the latter decomposes on heating (see question 1) and the +2 oxidation is generally more stable for lead.

In practice, prolonged heating in air at about 400 °C causes partial oxidation to dilead(II) lead(IV) oxide, Pb_3O_4, which decomposes again at higher temperatures:

$$6PbO(s) + O_2(g) \rightleftharpoons Pb_3O_4(s)$$

EXPERIMENT 11.3 Observation and deduction exercise 4

If your students have attempted the earlier observation and deduction exercises they will find this one relatively easy. Remind them that, in an examination, many marks can be obtained for detailed and accurate reporting of observations, even when no firm conclusions are drawn from them.

Suggested mark scheme

Specimen results are shown in the tables below, with a suggested mark allocation. Since you may choose to use only part of this experiment for assessment we have separated the skills for each part, but they may, of course, be combined.

Tests with Q (Results Tables 11.5 and 11.6)
Total marks for observation and recording = 13.
Total marks for inference and deduction = 12.

Total mark	13–11	10–8	7–6	5–4	3–2	1–0
Scaled mark	5	4	3	2	1	0

Tests with H and I (Results Table 11.7)
Total marks for observing and recording = 14.
Total marks for inference and deduction = 15 (including answers to questions).

Total mark	15–12	11–9	8–7	6–5	4–3	2–0
Scaled mark	5	4	3	2	1	0

Specimen results

Results Table 11.5
Tests on Q

Method	Observations	Inferences
1. A spatula-measure of the powder was heated in an ignition tube, gently at first and then more strongly	The red–orange solid initially darkened on heating (1) and a colourless gas was evolved (1) as the solid melted to a red liquid (1). On cooling the liquid solidified to a yellow solid (1)	The red–orange colour of Q suggests that it could be Pb_3O_4 (1) and that the resulting yellow solid is PbO (1)
A glowing splint was lowered into the tube	The splint relit as the solid melted (1)	O_2 produced (1), probably by the decomposition of Pb_3O_4 to leave PbO (1)
2. About 2 cm^3 of dilute nitric acid was added to half a spatula-measure of the powder in a test-tube. The mixture was gently heated	The orange solid turned to a dark brown solid on heating (1). The resulting brown solid settled to the bottom on standing with a colourless liquid above it (1)	The PbO part of Pb_3O_4 ($2PbO{\cdot}PbO_2$) dissolves (1) leaving brown PbO_2 (1)
3. The colourless liquid from test 2 was decanted off carefully into a test-tube		The colourless liquid is likely to contain $Pb^{2+}(aq)$ (1)
a About 1 cm^3 was poured into another test-tube and a few drops of KI solution was added	A yellow precipitate was formed (1)	Yellow precipitate of PbI_2 (1)
b NaOH(aq) was added to half the remaining solution from test 2, drop-by-drop initially and then to excess	White precipitate formed initially (1) but this redissolved on addition of excess NaOH(aq) (1)	White precipitate of $Pb(OH)_2$ formed (1) which dissolved to form plumbate(II) ions (1)

Results Table 11.6
Experiment to test inference

Inference	Test and observations	Conclusion
The resulting solution from test 2 contains $Pb^{2+}(aq)$	A yellow precipitate (1) appeared on adding a few drops of $CrO_4^{2-}(aq)$ to the unknown solution (1). The precipitate turned orange on heating (1)	Yellow $PbCrO_4$ is formed (1) confirming the presence of Pb^{2+} (1)

Results Table 11.7
Tests on H and I

Test	Observations	Inferences
1. To 1 cm^3 of the solution of H add aqueous silver nitrate followed by dilute nitric acid	White precipitate appeared on adding silver nitrate (1) with no further change on adding nitric acid (1)	AgCl(s) probably precipitated (1). The solution of H is likely to contain Cl^-(aq) ions (1)
2. To 1 cm^3 of the solution of H add aqueous sodium hydroxide until in excess	White precipitate appeared (1) which dissolved on addition of excess NaOH (1)	The precipitated hydroxide is amphoteric (1). Thus, any of the following metal ions could be present: Pb^{2+}, Zn^{2+}, Al^{3+}, Sb^{3+}, Sn^{2+} and Sn^{4+} (1)
3. To 1 cm^3 of aqueous iron(III) chloride add a few drops of aqueous potassium thiocyanate. To this solution add some of the solution of H	A red solution appeared on adding potassium thiocyanate to iron (III) chloride (1). The solution decolorised on addition of H (1)	This indicates that the Fe^{3+} has been reduced to Fe^{2+} by the solution of H (1). Therefore H is likely to contain Sn^{2+} since this is readily oxidised to Sn^{4+} (1)
4. Heat some of solid I in a Pyrex boiling-tube. Allow to cool. Add 8–10 cm^3 of dilute nitric acid to the residue and boil the mixture for one or two minutes. Filter if necessary and use portions of the cool solution for the following tests:	Dark brown solid melted on heating and gave off a colourless gas (1) which relit a glowing splint (1). On further heating the liquid became red (1) and cooled to a glassy yellow–brown solid (1)	Oxygen evolved (1) probably from the decomposition of brown PbO_2 to yellow PbO (1)
a To 1 cm^3 of the solution add aqueous sodium hydroxide	A white precipitate was formed (1) which dissolved on addition of excess NaOH (1)	Precipitated hydroxide is amphoteric. $Pb(OH)_2$ is probably precipitated which dissolves to give a plumbate(II) (1)
b To 1 cm^3 of the solution add dilute sulphuric acid	A white precipitate formed (1)	$PbSO_4$ probably precipitated (1)
c To 1 cm^3 of the solution add aqueous potassium chromate(VI)	Yellow precipitate formed (1)	$PbCrO_4$ precipitated (1) Confirms the presence of Pb^{2+} (1)

Answer to question

The metal contained in H is tin. (1)
In reaction 2 tin is in the +2 oxidation state. (1)
In reaction 3 tin is oxidised from +2 to +4 oxidation state. (1)

EXPERIMENT 11.4 Illustrating the oxidation states of vanadium

Zinc powder reduces vanadium(V) more rapidly than does granulated zinc and also seems to give less hydrogen (or perhaps the hydrogen is purer and therefore smells less!). Zinc **dust** is to be avoided since it is dangerously flammable and reacts too quickly.

Campden tablets (used for sterilising wine-making equipment and consisting largely of sodium disulphate(IV) – sodium metabisulphite – $Na_2S_2O_5$) may be used to reduce V(V) to V(IV).

Specimen results

Table 11.2

Ion (hydrated)	VO_2^+	VO^{2+}	V^{3+}	V^{2+}
Colour	Yellow	Blue	Green	Violet
Oxidation state	5	4	3	2
Name	Dioxovanadium(V)	Oxovanadium(IV)	Vanadium(III)	Vanadium(II)

Results Table 11.8

Test	**Observations**	**Summary of reaction**
Ammonium vanadate + acid	The white solid turned red and dissolved to a yellow solution	$\overset{+5}{VO_3^-} \rightarrow \overset{+5}{VO_2^+}$
Vanadium(V) + zinc	The zinc effervesced and the yellow solution became green, blue, green again and, eventually, violet	$\overset{+5}{VO_2^+} \rightarrow \overset{+4}{VO^{2+}} \rightarrow V^{3+} \rightarrow V^{2+}$
Vanadium(II) + manganate(VII)	The violet solution became green, blue, green–yellow and, finally, pink	$V^{2+} \rightarrow V^{3+} \rightarrow VO^{2+} \rightarrow VO_2^+$
Vanadium(V) + sulphite. Add vanadium(II)	The yellow solution became blue. On adding V(II) the mixture became green	$\overset{+5}{VO_2^+} \rightarrow \overset{+4}{VO^{2+}}$ $\overset{+4}{VO^{2+}} + V^{2+} \rightarrow V^{3+}$
Vanadium(V) + iodide + thiosulphate	The yellow solution became a muddy brown. Addition of thiosulphate gave a clear blue solution	$\overset{+5}{VO_2^+} \rightarrow \overset{+4}{VO^{2+}}$
Vanadium(II) + concentrated sulphuric acid	The violet solution became green	$V^{2+} \rightarrow V^{3+}$

Answers to questions

1. The first appearance of a green colour is due to a mixture of yellow vanadium(V) and blue vanadium(IV).
2. The subsequent changes are due to continuing reduction:

Blue → green V(IV) → V(III)
Green → violet V(III) → V(II) (slow)

These changes are reversed on adding manganate(VII) except that when oxidation is complete the solution is pink, due to an excess of MnO_4^- ions.

3. The addition of iodide ions turned the solution a muddy brown. Iodine was produced in the reaction.
4. Sodium thiosulphate was added to remove iodine and reveal the blue colour of vanadium(IV).
5. Since the zinc half-cell has the most negative electrode potential of those shown, it can perform all the reduction stages. But the electrode potentials show that iodide will reduce V(V) to V(IV) but not V(IV) to V(III):

$$2VO_2^+(aq) + 4H^+(aq) + 2I^-(aq) \rightleftharpoons 2VO^{2+}(aq) + 2H_2O(l) + I_2(aq)$$
$$\Delta E^\ominus = 1.00\ V - 0.54\ V = +0.46\ V$$

A positive value for $\Delta E^\ominus$= shows the reaction is feasible (from left to right as written).

$$2VO^{2+}(aq) + 4H^+(aq) + 2I^-(aq) \rightleftharpoons 2V^{3+}(aq) + 2H_2O(l) + I_2(aq)$$
$$\Delta E^\ominus = 0.34\ V - 0.54\ V = -0.20\ V$$

A negative value for $\Delta E^\ominus$= shows that this reaction is not feasible under standard conditions.

6. The addition of sulphite ions (which combine with acid to give H_2SO_3)* readily reduces V(V) to V(IV) giving a blue solution, but does not cause further reduction to V(III) despite the prediction from electrode potentials:

$$2VO^{2+}(aq) + 4H^+(aq) + H_2SO_3(aq) + H_2O(l) \rightleftharpoons 2V^{3+}(aq) + 2H_2O(l) + SO_4^{2-}(aq) + 4H^+(aq)$$

which, on cancelling $H^+(aq)$ and $H_2O(l)$, gives:

$$2VO^{2+}(aq) + H_2SO_3(aq) \rightleftharpoons 2V^{3+}(aq) + H_2O(l) + SO_4^{2-}(aq)$$
$$\Delta E^\ominus = 0.34\ V - 0.17\ V = +0.17\ V$$

Note that a greenish colour may appear if excess sulphite is added, but this is due to the alternative oxidation product $S_2O_3^{2-}$ which decomposes in acid solution to give sulphur. Filtering the greenish mixture gives a blue solution.

7. To oxidise V(II) to V(III) but not V(III) to V(IV) we need a mild oxidising agent with an electrode potential between −0.26 V and +0.34 V. Sulphuric acid (appearing in the table as $SO_4^{2-} + 4H^+$) with $\Delta E^\ominus = 0.17$ V should be suitable, and you can confirm that the careful addition of a few drops of concentrated sulphuric acid to a solution of V(II) does produce green V(III).

$$2V^{2+}(aq) + SO_4^{2-}(aq) + 4H^+(aq) \rightleftharpoons 2V^{3+}(aq) + H_2SO_3(aq) + H_2O(l)$$
$$\Delta E^\ominus = 0.17\ V - (-0.26\ V) = +0.43\ V$$

This helps to explain why the final stage in the reduction with zinc is rather slow. A fairly high concentration of sulphuric acid is needed in the first place to dissolve the ammonium trioxovanadate(V), but the final reduction of V(III) to V(II) will not occur until the zinc has lowered the concentration of acid somewhat by forming hydrogen.

*Equations may show $H_2SO_3(aq)$, $2H^+(aq) + SO_3^{2-}(aq)$, or $SO_2(aq) + H_2O(l)$.

EXPERIMENT 11.5 Illustrating the oxidation states of manganese

If time is short, this experiment could be omitted.

The alternative method mentioned for making Mn(V) involves reduction of Mn(VII) with 12 M KOH. In view of the hazards of handling very concentrated alkali, this is suggested as a teacher demonstration only. Hydroxide ions are oxidised to oxygen. Details can be found in *Nuffield Advanced Science: Chemistry, Teachers' Guide II* (First Edition), Section 16.1.

Answers to questions

1. $\Delta E^{\ominus}$ values lead us to expect formation of Mn(VI) only in alkaline solution. (Students may notice that the filter paper appears to act as a catalyst for this reaction.)
2. The addition of acid caused disproportionation of Mn(VI) to Mn(VII) (pink solution) and Mn(IV) (brown solid). The reduction in $[OH^-(aq)]$ caused the equilibrium to shift back to the left:

$$2MnO_4^-(aq) + MnO_2(s) + 4OH^-(aq) \rightleftharpoons 3MnO_4^{2-}(aq) + 2H_2O(l)$$

3. Dilution of the Mn(III) solution caused disproportionation by shifting the equilibrium back to the left:

$$4Mn^{2+}(aq) + MnO_4^-(aq) + 8H^+(aq) \rightleftharpoons 4H_2O(l) + 5Mn^{3+}(aq)$$

4. There is no red coloration.
5. The reactants are solids, which often react slowly.
6. The off-white precipitate of manganese(III) hydroxide darkens on standing, especially where it is in contact with the air. This fact suggests that manganese(II) hydroxide is readily oxidised by air to brown manganese(IV) oxide.

EXPERIMENT 11.6 Relative stabilities of some complex ions

Other ligands may be used if necessary – see, for instance, *Nuffield Advanced Science: Chemistry* (Third Edition), Section 15.4. Note that the addition of sodium ethanedioate gives a precipitate initially, which dissolves in excess.

Specimen results

Results Table 11.9

Ligand	H_2O		Cl^-		$C_2O_4^{2-}$ = ox		$NH_2C_2H_4NH_2$ = en		$C_{10}H_{12}O_8N_2^{4-}$ = edta	
Complex ion	$Cu(H_2O)_4^{2+}$		$CuCl_4^{2-}$		$Cu(ox)_2^{2-}$		$Cu(en)_2^{2+}$		$Cu(edta)^{2-}$	
Colour	Blue		Yellowish–green		Blue		Violet		Light blue	
Stability constant	–		4.0×10^5 $mol^{-4}\ dm^{12}$		2.1×10^{10} $mol^{-2}\ dm^6$		–		6.3×10^{18} $mol^{-1}\ dm^3$	
	Predictions (P) and results (R) ✓ = replacement, ✗ = none									
Test	**P**	**R**	**P**	**R**	**P**	**R**	**P**	**R**	**P**	**R**
Add H_2O			✗	✓	✗	✗		✗	✗	✗
Add Cl^-	✓	✓			✗	✗		✗	✗	✗
Add ox	✓	✓	✓	✓				✗	✗	✗
Add en	✓	✓	✓	✓	✓	✓			✗	✗
Add edta	✓	✓	✓	✓	✓	✓		✓		

Answers to questions

1. Yes. Generally, polydentate ligands give more stable complexes than monodentate ligands, stability increasing with the number of bonds formed. This is only a general rule, however, and it is possible to have a complex containing, say, strongly bonded monodentate ligands which is more stable than one containing weakly bonded bidentate ligands.
2. Since **en** ligands displace **ox** ligands, but not vice versa, and they are both bidentate, $Cu(en)_2^{2+}$ must have a substantially greater stability constant than $Cu(ox)_2^{2-}$. It is not directly comparable with the stability constant for $Cu(edta)^{2-}$ which has different units because edta is hexadentate.
3. H_2O and Cl^- ligands appear to be readily interchangeable when concentrations are changed.
4. $Cu^{2+}(aq) + 4Cl^-(aq) \rightleftharpoons CuCl_4^-(aq)$

$$K_{st} = \frac{[CuCl_4^{2-}(aq)]}{[Cu^{2+}(aq)][Cl^-(aq)]^4} = 4.0 \times 10^5\ mol^{-4}\ dm^{12}$$

$$\therefore \frac{[CuCl_4^{2-}(aq)]}{[Cu^{2+}(aq)]} = 4.0 \times 10^5\ mol^{-4}\ dm^{12} \times [Cl^-(aq)]^4$$

a $$\frac{[CuCl_4^{2-}(aq)]}{[Cu^{2+}(aq)]} = 4.0 \times 10^5 \times 5.0^4 = \mathbf{2.5 \times 10^8}$$

b $$\frac{[CuCl_4^{2-}(aq)]}{[Cu^{2+}(aq)]} = 4.0 \times 10^5 \times 0.05^4 = \mathbf{2.5}$$

These figures show that dilution of the solution will soon reduce $[Cl^-(aq)]$ sufficiently to convert a significant proportion of $CuCl_4^{2-}$ ions to $Cu(H_2O)_4^{2+}$ ions and so change the colour.

EXPERIMENT 11.7 Determining the formula of a complex ion

The experiment was tested using a WPA colorimeter (CO65). The concentrations given, and some of our instructions, might not necessarily be suitable for other colorimeters.

Adequate results can be obtained using no filters, but we think that the principles of filter selection are important enough to be included.

Specimen results

Results Table 11.10

Tube number	1	2	3	4	5	6	7	8
Volume of $(NH_4)_2SO_4/cm^3$	15	5	5	5	5	5	5	5
Volume of $CuSO_4/cm^3$	0.0	1.0	1.5	2.0	2.5	3.0	4.0	5.0
Volume of NH_3/cm^3	0.0	9.0	8.5	8.0	7.5	7.0	6.0	5.0
Colorimeter reading	0.00	0.20	0.29	0.34	0.32	0.28	0.22	0.15

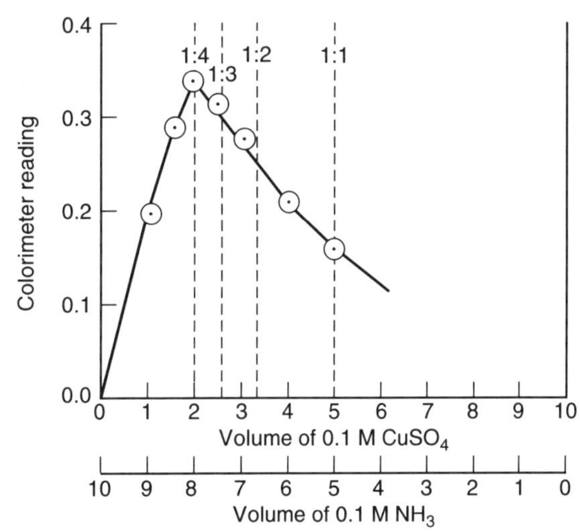

The maximum absorbance for a ratio $CuSO_4 : NH_3 = 1 : 4$ indicates that the formula is $Cu(NH_3)_4^{2+}$.

Answers to questions

1. The addition of ammonia solution to copper sulphate solution usually causes precipitation of pale blue copper hydroxide:

$$NH_3(aq) + H_2O(l) \rightleftharpoons NH_4^+(aq) + OH^-(aq)$$

and

$$Cu^{2+}(aq) + 2OH^-(aq) \rightarrow Cu(OH)_2(s)$$

However, the addition of ammonium sulphate reduces $[OH^-(aq)]$ so much, by the common ion effect, that precipitation does not occur.

2. Although the amount of copper present increases, there is less and less ammonia present to convert it into the highly-coloured complex ion. The concentration of the complex ion and, therefore, the absorbance, decreases.
3. Although the amount of ammonia decreases, there is always an excess up to the peak, so that the concentration of complex ion increases as the amount of copper present increases.
4. If filtering and absorbing occurred at single precise wavelengths, it would be vital to choose the correct filter. In practice, however, both filtering and absorbing occur over bands of wavelengths, and a certain amount of selective absorbance will occur

using any filter. Using the wrong filter in this experiment would still give the same final result although the peak would be less pronounced. More care may be necessary in other experiments where species of different colours with similar intensities may be present.

5. Since the colour of $Cu(NH_3)_4^{2+}(aq)$ is far more intense than that of $Cu(H_2O)_4^{2+}(aq)$, reasonable results can be obtained in this experiment by using no filter at all! Students were asked to use a filter to establish the general principle, which may be far more important in other similar experiments.

EXPERIMENT 11.8 Some redox chemistry of copper

Note that the use of boiling concentrated hydrochloric acid, often recommended for the preparation of copper(I) chloride, is not necessary. If copper(II) chloride is not available, a mixture of copper(II) sulphate and sodium chloride may be used.

Answers to questions

1. The white solid is copper(I) chloride, CuCl. Colour in the transition elements is associated with partly filled d orbitals. Cu(I) has a fully filled set of d orbitals – $1s^2 2s^2 2p^6 3s^2 3p^6 3d^{10}$.
2. In part A, the reducing agent is copper(0), in part B it is sulphur(IV).
3. Heat was required in part A because the reaction involved a solid – such reactions are often slow at room temperature.
4. The excess of chloride ions stabilises the Cu(I) oxidation state in a complex ion so that reduction of Cu(II) to Cu(I) can proceed in solution.
5. The dark brown colour is due to the increasing concentration of the copper(I) complex $CuCl_4^{3-}$ as it is formed from the green–yellow $CuCl_4^{2-}$.
6. $Cu^+(aq)$, present in very small concentration, is oxidised by air to $Cu^{2+}(aq)$. The equilibrium:

$$CuCl(s) \rightleftharpoons Cu^+(aq) + Cl^-(aq)$$

   is then shifted to the right to allow further oxidation of $Cu^+(aq)$ to $Cu^{2+}(aq)$, eventually giving a blue solution.
7. A muddy brownish suspension is formed. On settling, the precipitate appears creamy-coloured and the solution brown.
8. The thiosulphate removes the brown coloration, leaving a colourless solution over a creamy precipitate. The brown coloration must therefore have been due to iodine.
9. Since iodide has been oxidised to iodine, copper(II) must have been reduced to copper(I).
10. The precipitate must therefore be copper(I) iodide, CuI, a white solid made to appear creamy by traces of strongly adsorbed iodine.
11. The deep blue colour suggests the formation of a complex ion. The ligand has the formula $(CH(OH)CO_2^-)_2$ but we suggest that you do not need to learn the details – you can use the formula $Cu(lig)_2^{2-}$ for the complex ion.
12. The formation of complex ions often affects the ease with which redox reactions occur. Here, it is easier to reduce the copper(II) complex ion made using Fehling's solution than it is to reduce the aqua complex. Once the copper(I) complex is formed, there is sufficient ligand exchange with water to give a small concentration of $Cu^+(aq)$. This is prevented from disproportionating by the precipitation of red copper(I) oxide, Cu_2O. (You might well have predicted the precipitation of copper(I) hydroxide, CuOH, but this decomposes to the oxide and water even more readily than $Cu(OH)_2$, which requires heating.)

Results Table 11.11

	Method	Observations	Equation(s)
A Preparation of of copper(I) chloride	Heat copper(II) chloride, copper and excess sodium chloride in a little water. Dilute	The green solution darkened slowly to almost black. A white precipitate appeared on dilution	$CuCl_2(aq) + Cu(s) \rightarrow 2CuCl(s)$ (via $CuCl_4^{2-}$ and $CuCl_4^{3-}$)
B Preparation of of copper(I) chloride	Mix solutions of copper(II) chloride and sulphite ions	A white precipitate appeared	$2CuCl_2(aq) + SO_3^{2-}(aq) + H_2O(l) \rightarrow 2CuCl(s) + SO_4^{2-}(aq) + 2HCl(aq)$
C Preparation of of copper(I) iodide	Mix solutions of copper(II) sulphate and potassium iodide. Add sodium thiosulphate	A muddy brown suspension was formed. Adding thiosulphate left a cream precipitate and a colourless solution	$2Cu^{2+}(aq) + 4I^-(aq) \rightarrow 2CuI(s) + I_2(aq)$ $I_2(aq) + 2S_2O_3^{2-}(aq) \rightarrow 2I^-(aq) + S_4O_6^{2-}(aq)$
D Preparation of copper(I) oxide	Mix equal volumes of Fehling's solutions A and B. Add a little glucose and warm	A deep blue solution was formed on mixing A and B. On warming, a red solid was formed after intermediate muddy green and brown colours	$Cu(H_2O)_4^{2+} + 2lig^{2-} \rightleftharpoons Cu(lig)_2^{2-} + 4H_2O$ $Cu(lig)_2^{2-} + e^- \rightleftharpoons Cu(lig)_2^{3-}$ $Cu(lig)_2^{3-} + 4H_2O \rightleftharpoons Cu(H_2O)_4^{+} + 2lig^{2-}$ $2Cu^+(aq) + 2OH^-(aq) \rightleftharpoons Cu_2O(s) + H_2O(l)$

EXPERIMENT 11.9 Investigating the use of cobalt(II) ions as a catalyst

See notes for Experiment 11.10 below.

Answers to questions

1. Carbon dioxide from the redox reaction, and oxygen from the decomposition of hydrogen peroxide, which is also catalysed by Co^{2+}.
2. The dark green colour is due to a complex ion of Co(III) with 2,3-dihydroxybutanedioate ions as ligands. The green colour disappears as this complex completes the oxidation of excess 2,3-dihydroxybutanedioate and is itself reduced back to Co(II).
3. The green colour persists at room temperature because the reaction between the Co(III) complex and 2,3-dihydroxybutanedioate ion is slow.

EXPERIMENT 11.10 Catalysing the reaction between iodide ions and peroxodisulphate ions

You may feel that students need not attempt both Experiments 11.9 and 11.10. Experiment 11.9 is much simpler and shorter, but the chemistry in Experiment 11.10 is more important.

Specimen results

Results Table 11.12

Transition element		Chromium	Manganese	Iron
Oxidation states used		Cr(VI) Cr(III)	Mn(VII) Mn(II)	Fe(III) Fe(II)
Electrode potential/V (higher ox. state $\rightleftharpoons$ lower)		1.33	1.51	0.77
Electrode potential/V ($I_2 + 2e^- \rightleftharpoons 2I^-$)		0.54	0.54	0.54
Electrode potential/V ($S_2O_8^{2-} + 2e^- \rightleftharpoons 2SO_4^{2-}$)		2.01	2.01	2.01
Does higher ox. state oxidise I^-?	**Prediction**	Yes	Yes	Yes
	Practically	Yes	Yes	Yes
Does lower ox. state reduce $S_2O_8^{2-}$?	**Prediction**	Yes	Yes	Yes
	Practically	No	No	Yes
Do you expect catalysis?		No	No	Yes
Time for blue colour to appear *t*/s		Varies with conditions		

Answers to questions

1. Iron is an effective catalyst, both as Fe(II) and Fe(III). A simplified mechanism might consist of two steps, either of which could come first:

   Fe(II) reacts with peroxodisulphate $\rightarrow$ Fe(III) and sulphate
   Fe(III) reacts with iodide $\rightarrow$ Fe(II) and iodine

2. Chromium cannot function as a catalyst because Cr(III) will not react with peroxodisulphate. This is clearly a kinetic effect, because $\Delta E^{\ominus}$ is favourable.
3. Manganese(VII) functions well as a catalyst even though Mn(II) will not react with peroxodisulphate. Another oxidation state – Mn(III), Mn(IV) or Mn(VI) – could be produced as an intermediate.
4. Laboratory temperatures can vary considerably over two days (or even over a double period, or in different parts of the room!). As you have learned, temperature changes have a marked effect on rates of reaction and could make it difficult to decide whether catalysis has occurred.

EXPERIMENT 11.11 Observation and deduction exercise 5

Students may ask about the reaction in test 4**c** for which no inference was required. The transient deep-blue colour is due to the unstable peroxide which is usually given the formula CrO_5. Students may recall seeing this before in a test for hydrogen peroxide – see Experiment 4.1. The net result of the reaction is the reduction of Cr(VI) to Cr(III) and the **oxidation** of hydrogen peroxide to water and oxygen, as might be predicted from standard redox potentials.

Suggested mark scheme

Total marks for observing and recording = 30, as in Results Table 11.13 below.
Total marks for inference and deduction = 30, as in Results Table 11.13 and the oxidation states below.

Total mark	30–24	23–19	18–14	13–9	8–4	3–0
Scaled mark	5	4	3	2	1	0

Oxidation states of the metal in D:

4. **a** $+6 \rightarrow +3$
4. **b** $+6 \rightarrow +6 \rightarrow +6$
4. **d** $+3 \rightarrow +6$ (7)

Specimen results

Results Table 11.13

Test	Observations	Inferences
1. Warm three-quarters of your sample of C with 4–5 cm^3 of aqueous sodium hydroxide	The solid dissolved to a colourless solution (1). A pungent smell was noted on warming (1). The gas turned red litmus blue and gave white smoke in HCl fumes (1)	Ammonia released (1). C is an ammonium salt (1)
2. Add approximately 10 cm^3 of dilute sulphuric acid to the remainder of your sample of C in a boiling-tube and warm the mixture in order to dissolve the solid. Use portions of the solution for the following tests:	The white solid turned orange (1) and dissolved to a yellow solution (1)	C may contain a transition metal (1)
a To 1–2 cm^3 of the solution of C add an equal volume of aqueous potassium iodide. Then add, dropwise, aqueous sodium thiosulphate until there is no further change	The solution turned dark brown (1). Addition of $Na_2S_2O_3$ gave a blue solution (1)	Iodide oxidised to iodine (1) which was removed by $Na_2S_2O_3$ (1). Colour changes in C suggest the presence of a transition metal (1)
b To 2 cm^3 of the solution add a little copper powder and warm	The solution turned from yellow to green (1) and then to blue (1)	C is reduced by copper (1). Possibly vanadium changing from +5 to +4 (1)
c To 3–4 cm^3 of the solution add a little zinc powder and allow the mixture to stand. It is suggested that you make observations for about five minutes and then allow the mixture to stand for a further 30 minutes, making observations from time to time. During this time you should proceed with other tests	The zinc effervesced (1). The solution slowly turned green (1), then blue (1), then green again (1) and eventually violet (1). The gas popped when ignited (1)	Hydrogen evolved (1). C is reduced through several oxidation states (1) Probably vanadium +5 $\rightarrow$ +4 (blue) (1) $\rightarrow$ +3 (green) (1) $\rightarrow$ +2 (violet) (1)
3. Carry out a flame test on substance D	The flame turned lilac (1)	D is a potassium salt (1)
4. Dissolve some D in the minimum quantity of distilled water and use portions of the solution for the following tests:	An orange solution was formed (1)	D may contain a transition metal (1)
a Add a few drops of this solution to a mixture of 1–2 cm^3 of aqueous potassium iodide with an equal volume of dilute sulphuric acid. Then add aqueous sodium thiosulphate dropwise	The solution turned dark brown (1). Addition of $Na_2S_2O_3$ gave a blue–green solution (1)	Iodide oxidised to iodine (1) which was removed by $Na_2S_2O_3$ (1). Colour changes in D suggest the presence of a transition metal (1)
b To 1 cm^3 of the solution of D add aqueous sodium hydroxide. Then add dilute sulphuric acid until in excess	The solution turned yellow (1) and, on addition of acid, became orange (1)	D is a dichromate (1) which changes reversibly to a chromate (1)
c To about 2 cm^3 of the solution of D add an equal volume of dilute sulphuric acid. Then add hydrogen peroxide solution drop-wise until there is no further change	The first drop gave a deep blue colour (1). This disappeared (1) with effervescence (1) which increased with more H_2O_2 (1). The solution darkened and then became blue–green (1)	(No inference required)
d Transfer the solution from **c** to a boiling-tube and add about 10 cm^3 of aqueous sodium hydroxide and 1 cm^3 of hydrogen peroxide solution. Heat the resulting solution	A green ppt. (1) was formed which dissolved to a green solution (1). Warming with H_2O_2 gave a yellow solution (1) and a little bubbling (1)	D is reduced in acid solution by H_2O_2 (1) but oxidised again in alkali (1). Some H_2O_2 decomposed to oxygen (1)

EXPERIMENT 12.1 Reactions of aluminium

Since mercury(II) chloride is toxic, we have restricted its use to one or two drops. You may prefer to demonstrate this part, or omit it altogether. Copper(II) chloride removes the oxide layer just as effectively for reactions in solution but does not promote a reaction with oxygen, presumably because the oxide layer is re-formed. Note that chloride ions seem to be necessary for the removal of the oxide layer – other mercury(II) or copper(II) solutions are not effective.

Students need not write full equations, involving 6-co-ordinated complex ions, at this stage. The reactions are further elucidated later.

Specimen results

Results Table 12.1
Reaction of aluminium foil

Reagent	Observations	Identity of any gas given off
Sodium hydroxide solution	Slow evolution of gas at first, becoming quicker	Hydrogen
Sodium hydroxide solution after immersion in $CuCl_2(aq)$	Rapid evolution of gas	Hydrogen
Dilute hydrochloric acid	No reaction in cold. Gas evolved on heating	Hydrogen
Dilute hydrochloric acid after immersion in $CuCl_2(aq)$	Gas evolved in the cold	Hydrogen
Air after immersion in $CuCl_2(aq)$	No visible reaction	–
Air after immersion in $HgCl_2(aq)$	White crumbly solid appeared on surface	–

Results Table 12.2
Reactions of the solutions

Reagent	Observations using solutions of	
	Al in NaOH	Al in HCl
Dilute sulphuric acid	A white gelatinous precipitate appeared which dissolved on shaking. More acid made the ppt. permanent and still more redissolved it. Heat produced	
Sodium hydroxide solution		A white gelatinous precipitate appeared which dissolved on shaking. More alkali made the ppt. permanent and still more redissolved it. Heat produced
Sodium carbonate solution		A gas was evolved (CO_2) and a white gelatinous precipitate appeared which dissolved on shaking. More alkali made the ppt. permanent but it did not redissolve. Heat produced

Answers to questions

1. Aluminium is normally covered in a thin invisible layer of aluminium oxide which is not easily removed and protects the metal surface from further attack by air and even by dilute acids. Any oxide formed on the surface of iron does not adhere well enough to protect the metal. (Remember also that $E^{\ominus}$ values refer only to standard conditions.)
2. Solutions of copper(II) chloride and mercury(II) chloride remove the oxide layer and expose the aluminium surface, which then reacts more vigorously with acids and alkalis. (How the oxide layer is removed is not clear. It is interesting to note that the chloride ions play a part – other copper salts are not effective.)
3. Displacement reactions occur with the cations in solution, producing metallic copper and mercury. The copper is simply washed off and the aluminium surface then becomes again covered with a protective layer of oxide. On the other hand, liquid mercury dissolves some aluminium to form an alloy (called an amalgam) which exposes aluminium atoms to attack by oxygen but does not allow a coherent protective layer to form.
4. Washing soda and many oven-cleaners are alkaline in solution and therefore tend to attack aluminium pans and other kitchenware.
5. **a** For a transfer of six electrons:

$$2Al(s) \rightarrow 2Al^{3+}(aq) + 6e^-$$
$$6H^+(aq) + 6e^- \rightarrow 3H_2(g)$$

Adding: $2Al(s) + 6H^+(aq) \rightarrow 2Al^{3+}(aq) + 3H_2(g)$

b For a transfer of six electrons:

$$2Al(s) + 8OH^-(aq) \rightarrow 2Al(OH)_4^-(aq) + 6e^-$$
$$6H_2O(l) + 6e^- \rightarrow 3H_2(g) + 6OH^-(aq)$$

Adding: $2Al(s) + 2OH^-(aq) + 6H_2O(l) \rightarrow 2Al(OH)_4^-(aq) + 3H_2(g)$

6. The addition of acid first neutralises excess alkali and then removes hydroxide ions from the aluminate, allowing a precipitate of aluminium hydroxide to appear. The precipitate dissolves in excess acid.

$$Al(OH)_4^-(aq) + H^+(aq) \rightarrow Al(OH)_3(s) + H_2O(l)$$
$$Al(OH)_3(s) + 3H^+(aq) \rightarrow Al^{3+}(aq) + 3H_2O(l)$$

The addition of sodium hydroxide first neutralises excess acid, and then a precipitate of aluminium hydroxide appears. The precipitate dissolves in excess alkali to form aluminate ions.

$$Al^{3+}(aq) + 3OH^-(aq) \rightarrow Al(OH)_3(s)$$
$$Al(OH)_3(s) + OH^-(aq) \rightleftharpoons Al(OH)_4^-(aq)$$

The addition of sodium carbonate first neutralises excess acid, releasing carbon dioxide. Sodium carbonate is alkaline by hydrolysis:

$$CO_3^{2-}(aq) + H_2O(l) \rightleftharpoons HCO_3^-(aq) + OH^-(aq)$$

so that a precipitate of aluminium hydroxide again appears. However, the concentration of hydroxide ions is never sufficient to redissolve the precipitate.

EXPERIMENT 12.2 Reactions of the oxoacids of nitrogen and their salts

Ideally, the whole experiment should be performed at a fume cupboard. If space is limited, some of the tests may be done in a well-ventilated laboratory but test 1 must be done at a fume cupboard and on **a small scale** since NO_2 is toxic.

Specimen results

Results Table 12.3

Reactions of nitric acid and nitrates

Reactants		Observations	Identity of gas
1. Copper and	16 M HNO_3	Rapid evolution of red–brown gas. Tube became hot. Green solution formed	Nitrogen dioxide, NO_2
2. Copper and	8 M HNO_3	Slow bubbling. Gas was colourless near solution, red–brown near mouth of tube. Solution turned blue	Nitrogen monoxide, NO, then $NO \rightarrow NO_2$
3. Magnesium and	2 M HNO_3	Rapid bubbling. Gas was brownish but also popped with a lighted splint	Nitrogen dioxide and hydrogen, NO_2 and H_2
4. Magnesium and	0.5 M HNO_3	Slow bubbling. Gas was colourless and popped with a lighted splint	Hydrogen, H_2
5. Devarda's alloy and a nitrate in alkali		Rapid bubbling. Gas was colourless, popped with a lighted splint and turned red litmus paper blue	Hydrogen and ammonia, H_2 and NH_3
6. Aluminium and a nitrate in alkali		Slow bubbling at first, then faster. Gas was colourless, flammable and turned red litmus paper blue	Hydrogen and ammonia, H_2 and NH_3

Note that the reaction of nitric acid with a metal always gives a **mixture** of gaseous products, e.g. NH_3, N_2, N_2O, NO, NO_2, H_2 (and N_2H_4, NH_4^+). One product may be predominant under particular conditions. Mg appears to be the only metal to give H_2 but the 'pop' test may not always work in the presence of other gases.

Results Table 12.4
Reactions of nitrous acid and nitrites

Reactants in solution	**Observations**	**Identity of coloured product(s)**	**Oxidation or reduction of NO_2^-**
7. Sodium nitrite and sulphuric acid	Solution turned blue. Slow effervescence to give brownish gas	HNO_2 or perhaps N_2O_3. Then see (8) below	None
8. Warm nitrous acid	Blue colour disappeared. Rapid bubbling. Gas was colourless near solution, brown near mouth of tube	NO_2 (from $NO + O_2$)	Oxidation (and reduction – see question 4)
9. Potassium iodide and nitrous acid	Brown solution formed and black precipitate. Trace of brownish gas	Iodine (I_2 & I_3^-) NO_2	Reduction (as in 8)
10. Potassium manganate(VII) and nitrous acid	Solution became colourless. (Trace of brown precipitate which then cleared)	(MnO_2 due to local deficiency of acid)	Oxidation
11.**a** Iron(II) sulphate and sodium hydroxide	Green precipitate formed	$Fe(OH)_2$	None
11.**b** Iron(II) sulphate, nitrous acid and sodium hydroxide	Mixture turned black, then gave a yellow solution. Addition of alkali gave red–brown precipitate.	$Fe(H_2O)_5(NO)^{2+}$ Fe^{3+} $Fe(OH)_3$	None Reduction None
12. Aluminium and a nitrite in alkali	Rapid bubbling. Gas was colourless, popped with a lighted splint and turned red litmus blue	None	Reduction

Answers to questions

1. **Concentrated nitric acid**

$$Cu(s) + 4HNO_3(aq) \rightarrow Cu^{2+}(aq) + 2NO_3^{2-}(aq) + 2NO_2(g) + 2H_2O(l)$$

The red–brown gas is nitrogen dioxide, NO_2. The blue–green solution is concentrated aqueous copper nitrate.

Moderately concentrated nitric acid

$$3Cu(s) + 8HNO_3(aq) \rightarrow 3Cu^{2+}(aq) + 6NO_3^{2-}(aq) + 2NO(g) + 4H_2O(l)$$

The blue solution is dilute aqueous copper nitrate. The colourless gas is nitrogen monoxide, NO, which reacts with oxygen in the air to produce nitrogen dioxide, NO_2.

$$2NO(g) + O_2(g) \rightarrow 2NO_2(g)$$

2. The reaction between magnesium and very dilute nitric acid is a simple electron transfer between metal and hydrogen, as occurs with other dilute acids.

$$Mg(s) + 2H^+(aq) \rightarrow Mg^{2+}(aq) + H_2(g)$$

At higher concentrations of acid, the combination of nitrate and hydrogen ions becomes a more powerful oxidant than hydrogen ions alone and is reduced by the magnesium to give water and oxides of nitrogen.

3. The main reason is probably that the finely divided alloy presents a greater surface area for reaction than the aluminium foil. Another reason may be that the alloy is not protected initially by a coherent film of aluminium oxide.

4. $$3HNO_2(aq) \rightarrow H^+(aq) + NO_3^-(aq) + 2NO(g) + H_2O(l)$$

This reaction is an example of disproportionation. The oxidation state of nitrogen both increases and decreases, i.e. it changes from +3 to +5 and +2. Nitric acid cannot undergo disproportionation because the oxidation state of nitrogen is already at its maximum value of +5.

5. **a** Both nitrates and nitrites in alkaline solution react with aluminium or Devarda's alloy to produce ammonia. (Ammonium salts, of course, react similarly but can easily be recognised because ammonia is produced by the action of alkali alone.)
   b Nitrates do not react with dilute sulphuric acid. Nitrites do react to produce nitrogen monoxide, which immediately combines with oxygen to give red–brown fumes of nitrogen dioxide.

EXPERIMENT 12.3 Investigating reactions of the oxosalts of sulphur

We have included many more reactions than students would be expected to know and explain; we suggest that you emphasise that they are not expected to learn **all** the equations! These reactions provide a very useful exercise in the accurate recording of observations, although we feel the experiment as a whole is not straightforward enough to be used for practical assessment.

The silver nitrate tests may be omitted if you feel they are too extravagant – they are unlikely to appear in practical examinations but students should know about the dithiosulphatoargentate(I) complex in connection with its use in photography.

Note that the conversion of sulphite to thiosulphate is far from complete, so that X still gives the reactions of sulphite ions.

Specimen results

Results Table 12.5

Test	Na_2SO_3	Na_2SO_4	$Na_2S_2O_3$	$Na_2S_2O_8$	X
1. Dilute hydrochloric acid. Warm	**a** No reaction in cold. Bubbles of gas on warming. Choking smell. $K_2Cr_2O_7$ paper turned green (SO_2 produced)	**b** No visible reaction	**c** Solution turned slightly cloudy, denser when warm. Choking smell. $K_2Cr_2O_7$ paper turned green (SO_2 produced)	**d** No reaction in cold. Bubbles of gas on warming. Choking smell. Litmus paper bleached (Cl_2 produced)	Solution turned slightly cloudy, denser when warm. Choking smell. $K_2Cr_2O_7$ paper turned green (SO_2 produced)
2. Silver nitrate solution	**e** Initial white ppt. dissolved on shaking. With more $AgNO_3$ a dense white ppt. remained	**f** No reaction at first, then a faint white ppt. appeared	**g** Initial white ppt. dissolved on shaking. With more $AgNO_3$ the ppt. remained and turned yellow, brown and black	**h** Blackish ppt. formed slowly	Initial white ppt. dissolved on shaking. With more $AgNO_3$ the ppt. remained and turned yellow, brown and black
3. Iodine solution in aqueous potassium iodide	**i** The brown colour was immediately discharged. (Iodine reduced)	**j** No visible reaction	**k** The brown colour was immediately discharged. (Iodine reduced)	**l** The brown colour became darker. (Iodide oxidised)	The brown colour was immediately discharged. (Iodine reduced)
4. Potassium iodide solution	**m** No visible reaction	**n** No visible reaction	**o** No visible reaction	**p** A dark brown solution was formed (Iodide oxidised)	No visible reaction
5. Iron(III) chloride solution. Acidify, warm, add alkali	**q** A dark red–brown solution was formed which became almost colourless when hot. Addition of alkali gave a green ppt. (Fe^{3+} reduced)	**r** A yellow solution was formed which darkened a little on warming. Addition of alkali gave a red–brown ppt. (Fe^{3+} not reduced)	**s** A dark purple solution was formed which cleared when hot and then became cloudy. Addition of alkali gave a green ppt. (Fe^{3+} reduced)	**t** A yellow solution was formed which darkened a little on warming. Addition of alkali gave a red–brown ppt. (Fe^{3+} not reduced)	A dark red–brown solution was formed which became almost colourless when hot. Addition of alkali gave a green ppt. (Fe^{3+} reduced)
6. Effect of heat on solid	**u** Crystals turned white and gave off a steamy vapour which condensed on the upper tube (H_2O). White residue turned yellow on strong heating	**v** A colourless liquid was rapidly formed, which boiled to give a steamy vapour (H_2O) and a white residue. No further reaction	**w** A colourless liquid was rapidly formed which boiled to give a steamy vapour (H_2O).The yellowish residue turned brown and gave a black viscous liquid	**x** The solid melted to a colourless liquid. Bubbles of gas relit a glowing splint (O_2)	

Answers to questions

1. **a** Sulphur dioxide is released from 'sulphurous acid' on heating:

$$\overset{+4}{S}O_3^{2-}(aq) + 2H^+(aq) \rightarrow \overset{+4}{S}O_2(g) + H_2O(l)$$

b No reaction.

c Sulphur is slowly precipitated and sulphur dioxide is released:

$$\overset{+2}{S}_2O_3^{2-}(aq) + 2H^+(aq) \rightarrow \overset{+4}{S}O_2(g) + \overset{0}{S}(s) + H_2O(l)$$

d Sulphate ions are produced and chlorine is released:

$$\overset{+6}{S}_2O_8^{2-}(aq) + 2Cl^-(aq) \rightarrow 2\overset{+6}{S}O_4^{2-}(aq) + Cl_2(aq)$$

(Sulphur in $S_2O_8^{2-}$ **appears** to have an oxidation number of +7, given by $\frac{1}{2}(-2 + 16) = 7$. However, as the name suggests, the ion contains an O—O link, which means that two of the O atoms have an oxidation number of −1. It is **these** atoms that are reduced.)

e Silver sulphite is precipitated, but this dissolves initially in excess sulphite to form a complex ion.

i) $\overset{+4}{S}O_3^{2-}(aq) + 2Ag^+(aq) \rightarrow Ag_2\overset{+4}{S}O_3(s)$

ii) $Ag_2\overset{+4}{S}O_3(s) + \overset{+4}{S}O_3^{2-}(aq) \rightarrow 2[Ag\overset{+4}{S}O_3]^-$

f Silver sulphate is precipitated, but this is moderately soluble and so the precipitate is not heavy:

$$2Ag^+(aq) + SO_4^{2-}(aq) \rightarrow Ag_2SO_4(s)$$

g Silver thiosulphate is precipitated but this dissolves initially in excess thiosulphate to form a complex ion. The precipitate decomposes to black silver sulphide.

i) $\overset{+2}{S}_2O_3^{2-}(aq) + 2Ag^+(aq) \rightarrow Ag_2\overset{+2}{S}_2O_3(s)$

ii) $Ag_2\overset{+2}{S}_2O_3(s) + 3\overset{+2}{S}_2O_3^{2-}(aq) \rightarrow 2[Ag(\overset{+2}{S}_2O_3)_2]^{3-}$

iii) $Ag_2\overset{+2}{S}_2O_3(s) + H_2O(l) \rightarrow Ag_2\overset{-2}{S}(s) + 2H^+(aq) + \overset{+6}{S}O_4^{2-}(aq)$

h It is not clear what happens here. The reaction may be similar to **g** with an unstable complex ion producing silver sulphide.

i Iodine is reduced to iodide ions and sulphite ions are oxidised to sulphate ions:

$$I_2(aq) + \overset{+4}{S}O_3^{2-} + H_2O(l) \rightarrow 2I^-(aq) + \overset{+6}{S}O_4^{2-}(aq) + 2H^+(aq)$$

j No reaction.

k Iodine is reduced to iodide ions and thiosulphate ions are oxidised to tetrathionate ions. (Students should recall that this reaction is used in titrations.)

$$I_2(aq) + 2\overset{+2}{S}_2O_3^{2-}(aq) \rightarrow 2I^-(aq) + \overset{+2.5}{S}_4O_6^{2-}(aq)$$

(As students should recall, the overall oxidation number of 2.5 for S in $S_4O_6^{2-}$ arises because two S atoms have *Ox* = −1 and two have +6.)

l Iodide ions (from the potassium iodide) are oxidised to iodine and peroxodisulphate ions are reduced to sulphate ions:

$$2I^-(aq) + \overset{+6}{S}_2O_8^{2-}(aq) \rightarrow I_2(aq) + 2\overset{+6}{S}O_4^{2-}(aq)$$

(See note for **d**.)

m, **n** and **o** No reaction.

p The same reaction as in **l**.

q An unstable complex ion of uncertain composition is formed initially but this decomposes. Iron(III) ions are reduced to iron(II) ions and sulphite ions are oxidised to sulphate ions. The green precipitate is iron(II) hydroxide:

$$2Fe^{3+}(aq) + \overset{+4}{S}O_3^{2-}(aq) + H_2O(l) \rightarrow 2Fe^{2+}(aq) + \overset{+6}{S}O_4^{2-}(aq) + 2H^+(aq)$$
$$Fe^{2+}(aq) + 2OH^-(aq) \rightarrow Fe(OH)_2(s)$$

r and **t** No reaction initially. The red–brown precipitate is iron(III) hydroxide:

$$Fe^{3+}(aq) + 3OH^-(aq) \rightarrow Fe(OH)_3(s)$$

s An unstable complex ion of uncertain composition is formed initially. Iron(III) ions are reduced to iron(II) ions and thiosulphate ions are oxidised to tetrathionate ions. The green precipitate is iron(II) hydroxide:

$$2Fe^{3+}(aq) + 2\overset{+2}{S}_2O_3^{2-}(aq) \rightarrow 2Fe^{2+}(aq) + \overset{+2.5}{S}_4O_6^{2-}(aq)$$
$$Fe^{2+}(aq) + 2OH^-(aq) \rightarrow Fe(OH)_2(s)$$

u Water of crystallisation is released. There is possibly some slight decomposition to sulphur (and sulphur dioxide?).

v Water of crystallisation is released rapidly, in sufficient quantity to dissolve the salt. The anhydrous sulphate does not decompose at the temperature of a Bunsen burner.

w Water of crystallisation is released rapidly, sufficient to dissolve the salt. The anhydrous thiosulphate decomposes, releasing sulphur, which melts to a thick black liquid.

x The peroxodisulphate releases oxygen on heating and forms the sulphate and sulphur dioxide:

$$\overset{+6}{S}_2O_8^{2-}(aq) \rightarrow \overset{+6}{S}O_4^{2-}(aq) + O_2(g) + \overset{+4}{S}O_2(g)$$ (See note for **d**.)

2. Reaction **c**, between thiosulphate and hydrogen ions, is disproportionation. [Also **g** (iii) and **w**.]
3. i) The peroxodisulphate ion is the strongest oxidising agent.
   ii) The sulphite ion and the thiosulphate ion are strong reducing agents. [Thiosulphate is stronger, under standard conditions, but this is not evident in these reactions.]
   iii) The sulphate ion is the most stable – it remains intact throughout these tests.
4. Dissolve the two salts in water. Add dilute hydrochloric acid and barium chloride solution to each. The sulphate gives a dense white precipitate of barium sulphate:

$$Ba^{2+}(aq) + SO_4^{2-}(aq) \rightarrow BaSO_4(s)$$

Barium sulphite is soluble in dilute acids but there may be a faint precipitate due to sulphate impurities formed by oxidation.

5. The reactions of X suggest that thiosulphate ions are formed by heating sulphur and sulphite ions:

$$S(s) + SO_3^{2-}(aq) \rightarrow S_2O_3^{2-}(aq)$$

The reaction takes some time to complete, so the prepared solution still contained some sulphite.

EXPERIMENT 12.4 Anodising aluminium

If you decide to omit this experiment to save time, we suggest that students should read the instructions so that they know how it is done. A larger cathode shaped as a cylinder may be used to surround the anode but the process is well illustrated with the simple apparatus we have described.

Answers to questions

1. Anodised aluminium is:
   a less shiny than non-anodised aluminium,
   b a poor conductor on the surface,
   c easily dyed.
2. The anodised layer adds some resistance to the circuit. The resistance of the solution increases as the temperature falls because the ions move less readily.
3. **a** Anodised aluminium is used to increase resistance to corrosion, e.g. in window frames, and for articles made more attractive to the consumer by dyeing, e.g. saucepan lids.
   b Anodised aluminium is not used for articles where electrical contacts need to be made.

EXPERIMENT 12.5 Observation and deduction exercise 6

Remind your students to look at the wording of the question. In **this** case, comments are required on the **types** of reaction involved.

Suggested mark scheme

Specimen results, with a total allocation of 50 marks, are given in Results Table 12.6 below. However, if you are using this test as a practical assessment, we suggest that scaled marks could be allocated as follows.

Total marks for observing and recording = 20, as in Results Table 12.6.

Total mark	20–17	16–14	13–11	10–8	7–4	3–0
Scaled mark	5	4	3	2	1	0

Total marks for inference and deduction = 25, as in Results Table 12.6.

Total mark	25–20	19–16	15–12	11–8	7–4	3–0
Scaled mark	5	4	3	2	1	0

Clearly, 5 marks in the 'Method' column cannot form the basis for an assessment of planning skills.

Specimen results

Results Table 12.6

Test	Observations	Inferences
1. To 2 or 3 cm^3 of solution B add a few drops of dilute hydrochloric acid. Now add a few drops of aqueous barium chloride	A dense white ppt. was formed on adding the barium chloride (1)	B probably a sulphate (1). Ba^{2+} combines with $SO_4^{2-} \rightarrow BaSO_4(s)$ (1)
2. To 2 or 3 cm^3 of solution B add a few drops of aqueous lead ethanoate (lead acetate). Now add an excess of aqueous ammonium ethanoate (ammonium acetate) and shake the mixture	A dense white ppt. was formed (1). It dissolved in excess ethanoate to a colourless solution (1)	Ppt. probably $PbSO_4$ (1) which is converted to soluble, covalent, lead(II) ethanoate (1)
3. To about 5 cm^3 of solution C in a boiling-tube add about 10 cm^3 of dilute hydrochloric acid. Allow the mixture to stand for a minute or two. Cautiously smell the mixture. Warm, if necessary, and test the gas evolved. Describe below how you performed this test	A white ppt. developed slowly (1). A pungent smell was noted on warming (1)	White ppt. probably S (1). Disproportionation (1) of thiosulphate (1) gives S and SO_2 (1)
Method A test-strip was soaked in acidified aqueous dichromate and held at the tube's mouth (2)	The paper changed from orange to green (1)	SO_2 confirmed (1). It reduces $Cr_2O_7^{2-}$ to Cr^{3+} (1)
4. **a** In a small beaker place 2 or 3 cm^3 of aqueous copper(II) sulphate. Acidify with two drops of dilute sulphuric acid. Now add about 5 cm^3 of aqueous potassium iodide	A cream ppt. (1) began to settle from a brown solution (1)	Cu^{2+} oxidises (1) I^- to I_2 and is reduced to CuI (1). Brown solution is $I_3^- - I_2$ in I^-(aq) (1)
b To the mixture obtained in 4 **a** add solution C until in excess, swirling well	Cream ppt. remained (1) in colourless solution (1)	Thiosulphate reduces I_2 to I^- (1)
5. **a** In a small beaker mix about 2 cm^3 of aqueous sodium chloride with an equal volume of aqueous silver nitrate	A dense white ppt. was formed (1)	AgCl ppt. from precipitation reaction between Ag^+(aq) and Cl^-(aq) (1)
b To the mixture obtained in 5 **a** add solution C until in excess, swirling well	The ppt. dissolved (1) to colourless solution (1)	$S_2O_3^{2-}$ and Ag^+ form soluble complex ion (1)
6. Dissolve about 1 g of iron(II) sulphate crystals in dilute sulphuric acid and add a few drops of aqueous potassium thiocyanate. Now add a few drops of solution D	No reaction at first (1) but addition of D gave a deep blood-red coloration (1)	Fe^{2+} oxidised (1) by D to Fe^{3+} which gives a red complex ion (1) with thiocyanate
7. **a** In a small beaker place about 5 cm^3 of aqueous potassium iodide and a few drops of dilute sulphuric acid. Now add about 5 cm^3 of solution D. Warm this mixture	A pale yellow solution (1) was formed which darkened to brown on warming (1)	D oxidises I^- to I_2 (1). Since D contains S and O it is probably a peroxodisulphate (1)
b To the mixture obtained in 7 **a** add solution C until in excess, swirling well	Solution became colourless (1) Faint white ppt. (1)	Thiosulphate reduces iodine to to iodide (1)
8. Evaporate a small quantity of solution B to dryness and perform a flame test on the residue. Describe below how you do this **Method** A Pt wire was cleaned in conc. HCl and held in a blue flame until no coloration was given. It was then dipped in B moistened with conc. HCl and again held in the flame (3)	A persistent orange–yellow flame colour was seen (1)	Sodium is present in B and also in C and D (1). From the conclusions noted above: B = Na_2SO_4 (1) C = $Na_2S_2O_3$ (1) D = $Na_2S_2O_8$ (1)

INTRODUCTION TO TECHNICIAN'S SHEETS

The laboratory technician plays a vital role in the smooth running of the experiments in *Advanced Practical Chemistry*. These Technician's Sheets are designed to help you organise the materials required in such a way that experiments can be completed successfully, safely and quickly.

Requirements

Each experiment has a Technician's Sheet which lists the requirements for one student (or pair) in four sections – glassware, other apparatus, pure chemicals and solutions – in alphabetical order within each section. The teacher should tell you how many students are likely to do the experiment in each session, and how apparatus will be shared, so that you can easily work out the total requirements from the list.

Trays We recommend that all the requirements for one experiment should be set out in a labelled tray (or trays), which can be reissued over several sessions if necessary.

Solutions We give instructions for making up all solutions. In some cases, specific instructions are given on the sheet, but there is also a general section summarising the preparation of solutions used in chapters 8 and 10 below. In general, the concentrations required are approximate, so that masses and volumes need not be measured with great accuracy – measuring cylinders and a balance of sensitivity ± 0.1 g are usually adequate. Where greater precision is required, we make it clear in the instructions. All aqueous solutions should, of course, be made up with distilled or deionised water.

Quantities We list the minimum quantity of each chemical that each student (or pair) will need. There is usually no need to measure out this quantity; it simply enables you to check that you always supply sufficient for the students' needs.

Chemical names We use systematic names (ASE) for chemicals throughout. However, since manufacturers' bottles are frequently labelled with traditional names, we give these as well (in brackets) where appropriate.

Planning ahead Please check through the Technician's Sheets for all the experiments so that items not in stock can be ordered or made in good time.

Preparation of solutions

As there are so many different solutions to make up for the experiments in chapters 8 and 10, some of the information is summarised below.

Ammonia solution, 2 M NH_3 To prepare 1 litre, measure 115 cm^3 of concentrated ammonia and add about 300 cm^3 of distilled water, stir, dilute to the final volume and stir again. Take great care opening bottles of 0.880 ammonia, particularly on hot days. A **face-shield** is recommended.

Bromine water This should be prepared in a fume cupboard. Use 1 cm^3 of bromine in 125 cm^3 of water. The safest procedure is to crush an ampoule under water using pliers. Store the solution in an amber bottle in a fume cupboard.

2,4-Dinitrophenylhydrazine solution Dissolve 0.5 g of solid in 25 cm^3 of methanol and add, dropwise, 1 cm^3 of concentrated sulphuric acid. Filter off any undissolved solid.

Fehling's solution 1 Dissolve 7 g of copper(II) sulphate crystals in water containing a few drops of dilute sulphuric acid, and make up to 100 cm^3.

Fehling's solution 2 Dissolve 12 g of sodium hydroxide pellets and 34 g of sodium potassium 2,3-dihydroxybutanedioate (tartrate), 'Rochelle salt', in water. Filter if necessary and make up to 100 cm^3.

Iodine solution, 10% I_2 in KI Dissolve 10 g of iodine and 20 g of potassium iodide in water and make up to 100 cm^3.

Iodine solution, 0.1 M I_2 in KI(aq) Dissolve 4 g of potassium iodide, KI, in 20 cm^3 of distilled water, add 2.5 g of iodine and stir to dissolve. Dilute to 100 cm^3.

Lime-water Shake 75 g of calcium hydroxide in 2 dm^3 of water in a Winchester bottle and allow to settle. Siphon off as required and refill with water.

Potassium dichromate solution, 0.1 M $K_2Cr_2O_7$ Dissolve 2.9 g of solid in distilled water and make up to 100 cm^3.

Potassium manganate(VII) (permanganate), 0.01 M $KMnO_4$ Solutions do not keep well unless the container is scrupulously clean and kept in an amber bottle. Add 1.6 g of solid to 60 cm^3. Stir, warm gently. When no more crystals remain make up to 100 cm^3 of water and stir again. Wear gloves and eye protection.

Silver nitrate solution, 0.05 M $AgNO_3$ Dissolve 2.1 g of solid in distilled water and make up to 250 cm^3. Store and dispense in amber bottles.

Sodium hydroxide solution 2 M NaOH Weigh out 80 g of sodium hydroxide pellets and add slowly, while stirring, to about 600 cm^3 of water in a beaker. Stir until dissolved then make up to 1 litre and stir again. Care should be taken that overheating does not make the solution boil over when dissolving the solid in water. Pellets should be used and should be stirred to prevent clumping as they dissolve. **Wear gloves and eye protection.** A choking mist is often formed as the solid dissolves in water. While this is not a serious safety risk, it is unpleasant and it is wise to use a fume cupboard where possible.

Starch solution, 2% Mix 2 g of **soluble** starch to a paste with a little distilled water. Add the paste to 100 cm^3 of boiling distilled water, with stirring, and boil for one minute. Cool. Keep for no longer than two weeks.

Sulphuric acid, dilute, 1 M H_2SO_4 Usual stock is very corrosive. Gloves and eye protection should be worn. Always add the acid to the water and do so with care. To prepare 1 litre, measure out 54 cm^3 of concentrated acid (98% w/v) in a measuring cylinder and put 600 cm^3 of distilled water in a 2-litre beaker. Pour the acid into the water in a thin stream while stirring with a glass rod using the other hand. Pour this solution into a 1-litre measuring cylinder and make up to the mark with distilled water.

■ Dispensing chemicals

You may already have a system for dispensing chemicals of different types in convenient containers and safe quantities. If not, we suggest the following.

Solids Either issue manufacturers' bottles (unless these are too large for convenience or safety) or dispense into small, clearly labelled screw-top jars.

Liquids Organic liquids especially are best dispensed in manufacturers' bottles (unless these are too large for convenience or safety). To help students to handle these liquids safely, without spillage, and without contaminating your stock, each bottle should be accompanied by a **labelled** teat pipette standing in its own small **labelled** beaker.

Aqueous solutions Except when large quantities are required, e.g. for titrations, solutions are best dispensed in small dropping bottles. If standard reagent bottles are used, teat pipettes should be issued together with two small beakers to enable students to wash and rinse them in distilled water before each use.

Safety Take careful note of hazard symbols printed on manufacturers' labels. If your stock bottles do not have modern labels, look at the requirements lists where we print hazard symbols and some specific hazard warnings. More details can be found in laboratory manuals and on HAZCARDS (see below).

Hazards The School Science Service (formerly the CLEAPSS development group) has produced a set of cards which not only state the potential dangers of commonly used chemicals but also tell you what to do in the event of an accident, and how to dispose of residues. It is a good idea to issue a HAZCARD with each dangerous chemical. These cards can be purchased from the Association for Science Education or from the School Science Service, Brunel University, Uxbridge UB8 3PH.

Disposal of organic residues

Contaminated residues may be disposed of as follows.

- **Up to 10 cm^3 at a time** Small quantities may be poured down a sink in a fume cupboard with plenty of flowing water and the addition of some detergent. If detergent is not added to some of the oily organic residues they may stick to the sides of the sink where they can then give off harmful vapour into the room. Residues from different experiments, containing different chemicals should not be mixed.
- **Larger volumes** You will need to consult your local education authority on the facility available for disposing of larger volumes, especially halogenoalkanes. Highly flammable solvents should **not** be put down the sewer in large amounts.
- **Organic solids** Small amounts can be put in the waste. If they are toxic then dilution of the solid with sand will render the mixture unpalatable (the authorities are concerned that children may put them in their mouths).

It is always advisable to consult your local education authority or water authority. Further details can be found in laboratory manuals such as CLEAPSS Laboratory Handbook. A full list of recommended books is given below.

Recommended books

These Technician's Sheets are not intended to provide a complete guide to laboratory management and safety. If you do not already have a reliable modern manual, we suggest you consider purchasing one or more of the following:

J. Creedy, *A Laboratory Manual for Schools and Colleges* (Heinemann) ISBN 0 435 57130 3.

K. Everett and E. Jenkins, *A Safety Handbook for Science Teachers* (John Murray) ISBN 0 7195 3760 6.

S. G. Luxon, *Hazards in the Chemical Laboratory* (Royal Society of Chemistry) ISBN 0 85186 2292.

ASE (The Association for Science Education), *Chemical Nomenclature and SI Units* (ASE, College Lane, Hatfield, Herts.) – a single publication replacing the two earlier booklets *Chemical Nomenclature, Symbols and Terminology* and *SI Units*.

ASE (The Association for Science Education), *Topics in Safety* ISBN 0863 571042 (ASE, College Lane, Hatfield, Herts.)

Safeguards in the School Laboratory (John Murray) ISBN 0863 570836 (new ed. 1996).

CLEAPSS, *Laboratory Handbook,*(School Science Service, Brunel University).

The following CLEAPSS guides also have a safety content: L195 (*Substitute Chemicals*) and L196 (*Risk Assessment for Science*).

EXPERIMENT 1.1 Determining the Avogadro constant

Requirements per student (or pair)	Notes
■ measuring cylinder, 10 cm^3	
■ teat pipette drawn out to a fine point or teat pipette with adaptor	adaptors called 'microtips' can be obtained from Hughes and Hughes Ltd The teat pipette should be capable of delivering about 100 drops to 1 cm^3
■ large trough – glass or earthenware	
■ safety spectacles	
■ fine cotton thread, 40–50 cm	
■ scissors	
■ petroleum jelly or vaseline	
■ oleic acid solution in pentane (0.005%) (in corked test-tube)	about 5 cm^3 0.50 cm^3 of oleic acid made up to 50.0 cm^3 with pentane and then 0.50 cm^3 of this solution made up to 100.0 cm^3 with pentane for the final solution If pentane is not available, petroleum ether (40–60, or even 30–40) can be used

HAZARD WARNING Pentane is extremely flammable and harmful by inhalation. Therefore you **must**:

- **keep the stopper on the bottle when not in use;**
- **keep the liquid away from flames.**

TECHNICIAN'S SHEET

EXPERIMENT 1.2 Preparing a standard solution

Requirements per student (or pair)

- beaker, 250 cm^3
- dropping pipette (teat pipette)
- filter funnel
- stirring rod with rubber end
- volumetric flask, 250 cm^3, with stopper
- weighing bottle
- spatula
- label for volumetric flask
- wash-bottle of distilled water
- safety spectacles
- access to balance capable of weighing to 0.01 g
- potassium hydrogenphthalate $C_8H_5O_4K$, AR

Notes

Potassium hydrogenphthalate: about 2 g. The students will be preparing a standard solution of potassium hydrogenphthalate which will be set aside in a labelled 250 cm^3 volumetric flask for Experiment 1.3

EXPERIMENT 1.3 An acid–base titration

Requirements per student (or pair)	Notes
■ 2 beakers, 100 cm^3	
■ burette, 50 cm^3	
■ 4 conical flasks, 250 cm^3	
■ filter funnel, small	to fill burette
■ pipette, 25 cm^3	
■ pipette filler	
■ burette stand	
■ safety spectacles	
■ white tile	
■ wash-bottle of distilled water	
■ standard potassium hydrogenphthalate solution	prepared by the students and kept from Experiment 1.2
■ sodium hydroxide solution ~ 0.1 M NaOH	about 150 cm^3. See below
■ phenolphthalein indicator solution	see below

HAZARD WARNING Sodium hydroxide solution is very corrosive. Even when dilute it can damage your eyes. Therefore you **must**:

■ **wear safety spectacles throughout the preparation.**

Approx. 0.1 M sodium hydroxide solution Dissolve 4.10 g of sodium hydroxide in freshly boiled and cooled distilled water. Make up 1000 cm^3 and invert 20 times to ensure thorough mixing.

Phenolphthalein indicator solution Dissolve 1 g of the reagent in 100 cm^3 of ethanol (IMS) and add 100 cm^3 of distilled water with constant stirring. Filter if a precipitate forms. Alternatively, the indicator solution can be purchased already prepared for use.

EXPERIMENT 1.4 A redox titration

Requirements per student (or pair)	Notes
■ 2 beakers, 100 cm^3	
■ burette, 50 cm^3	
■ 4 conical flasks, 250 cm^3	
■ filter funnel, small	to fill burette
■ pipette, 10 cm^3	
■ pipette filler	
■ burette stand	
■ safety spectacles	
■ white tile	
■ wash-bottle of distilled water	
■ iodine solution, ~ 0.05 M I_2 (standardised)	about 60 cm^3. See below
■ sodium thiosulphate solution, 0.0500 M $Na_2S_2O_3$	about 120 cm^3. See below
■ starch indicator solution	see below

Since the accuracy of the students' results depends on the concentrations of the iodine and sodium thiosulphate solutions, these **must** be prepared as carefully as possible.

Iodine solution 0.05 M I_2 Dissolve 12.7 g of iodine crystals and 20 g of potassium iodide crystals in distilled water and make up to 1000 cm^3. This should be standardised against the sodium thiosulphate solution.

Sodium thiosulphate solution, 0.0500 M $Na_2S_2O_3$ Dissolve 12.41 g of AR sodium thiosulphate, $Na_2S_2O_3{\cdot}5H_2O$ in distilled water and make up to 1000 cm^3. If the solution is to be kept for more than a few days, add 0.10 g of sodium carbonate, Na_2CO_3.

Starch indicator solution Make a paste of 1 g of soluble starch with a little water, pour the paste, with constant stirring, into 100 cm^3 of boiling water, and boil for about 1 minute. Allow to cool and add about 2.0 g of potassium iodide. Keep the solution in a stoppered bottle. Test it before the experiment and renew it when necessary.

EXPERIMENT 1.5 A precipitation titration

Requirements per student (or pair)	Notes
■ 2 beakers, 100 cm^3	
■ beaker, 250 cm^3	
■ burette, 50 cm^3	
■ 4 conical flasks, 250 cm^3	
■ dropping pipette (teat pipette)	
■ filter funnel, small	
■ pipette, 10 cm^3	
■ stirring rod with rubber end	
■ volumetric flask, 250 cm^3	
■ weighing bottle	
■ burette stand	
■ label for flask	
■ pipette filler	
■ safety spectacles	
■ spatula	
■ wash-bottle of distilled water	
■ white tile	
■ access to balance capable of weighing to 0.01 g	
■ access to large 'Silver residues' bottle	recovery of silver from the silver residues should be done every year or two, and silver nitrate regenerated
■ barium chloride crystals, $BaCl_2{\cdot}2H_2O$	students must see the name only, **not** the formula
■ sodium sulphate crystals	
■ silver nitrate solution, 0.0500 M $AgNO_3$	about 150 cm^3. See below
■ potassium chromate indicator solution	12.50 g of K_2CrO_4 made up to 250 cm^3 with distilled water

HAZARD WARNING

Barium chloride is harmful if swallowed. Silver nitrate is corrosive and can stain the skin. Therefore you **must**:
- **use the pipette filler provided;**
- **wash your hands after use.**

Silver nitrate solution 0.0500 M $AgNO_3$ 8.49 g of AR $AgNO_3$ made up to 1000 cm^3 with distilled water and stored in an amber bottle.

If only commercial recrystallised silver nitrate is available (not AR) the prepared solution must be standardised with sodium chloride solution using potassium chromate indicator.

EXPERIMENT 1.6 A titration exercise

Requirements per student	Notes
■ burette, 50 cm^3	
■ 3 conical flasks, 250 cm^3	
■ filter funnel, small	to fill burette
■ pipette, 25 cm^3	
■ burette stand	
■ pipette filler	
■ safety spectacles	
■ wash-bottle of distilled water	
■ white tile	
■ dilute sulphuric acid ~1 M H_2SO_4	if preparing by diluting the concentrated acid, add the acid **slowly** to the water with stirring. Use 54 cm^3 of concentrated acid to make 1 dm^3 of 1 M H_2SO_4
■ solution A of ammonium iron(II) sulphate, $FeSO_4(NH_4)_2SO_4 \cdot 6H_2O$	about 150 cm^3. See below
■ solution B of potassium manganate(VII), $KMnO_4$ (potassium permanganate)	about 150 cm^3. See below

HAZARD WARNING **Concentrated sulphuric acid** can cause severe burns and reacts vigorously with water. Therefore you **must:**

■ **wear gloves and eye protection and follow instructions for dilution carefully.**

Solution A. 39.2 g of ammonium iron(II) sulphate, $FeSO_4 \cdot (NH_4)_2SO_4 \cdot 6H_2O$ made up to 1000 cm^3 with distilled water and given the label: '**A – ammonium iron(II) sulphate, $FeSO_4(NH_4)_2SO_4 \cdot xH_2O$, 39.2 g dm^{-3}**'.

Solution B. 3.16 g of potassium manganate(VII) (permanganate) $KMnO_4$ made up to 1000 cm^3 with distilled water and then standardised using AR sodium ethandioate (oxalate) $Na_2C_2O_4$, to establish its precise concentration (close to 0.02 M).

This solution should be labelled as '**B – potassium manganate(VII), $KMnO_4$**' with the precise concentration in mol dm^{-3} given to the students.

EXPERIMENT 1.7 Using a hand spectroscope to observe emission spectra

Requirements per student (or pair)	Notes
■ beaker, 10 cm^3	
■ 2 watch glasses	
■ Bunsen burner and bench mat	
■ flame test wire (or tongs and filter papers)	preferably platinum
■ lamp, tungsten filament, pearl bulb	room fixture is adequate
■ pestle and mortar, small	
■ safety spectacles	
■ spatula	
■ spectroscope, hand held	
■ access to fume cupboard	
■ coloured plate of emission spectra	consult the teacher
■ barium chloride, $BaCl_2$	a few grams of each solid (barium chloride to strontium chloride)
■ calcium chloride, $CaCl_2$	
■ lithium chloride, LiCl	
■ potassium chloride, KCl	
■ sodium chloride, NaCl	
■ strontium chloride, $SrCl_2$	
■ hydrochloric acid, concentrated, HCl	

HAZARD WARNING

- **Hydrochloric acid** has a corrosive vapour which irritates and can damage your eyes. **Wear safety spectacles and work in the fume cupboard for this experiment.**
- ***Under no circumstances*** must you look through the spectroscope at the sun. If you do so, your eyes may be permanently damaged.

ILPAC 2

CHEMICAL ENERGETICS

TECHNICIAN'S SHEET

EXPERIMENT 2.1 Determining an enthalpy change of reaction

Requirements per student (or pair)	Notes
■ pipette, 25 cm^3	
■ weighing bottle	
■ balance (nearest 0.1 g)	
■ pipette filler	
■ polystyrene cup with lid	**expanded** polystyrene to minimise heat loss, or two thin cups one inside the other. Lid needs hole for thermometer
■ safety spectacles	
■ spatula	
■ stopclock	not needed if clock with second hand is visible in laboratory
■ thermometer, 0–100 °C	preferably graduated in 0.2 °C
■ copper sulphate solution, 1.00 M $CuSO_4$	about 30 cm^3. Dissolve 250 g of $CuSO_4{\cdot}5H_2O$ in distilled water and make up to 1000 cm^3
■ zinc powder	about 6 g. Granulated zinc is **not** suitable – reaction is too slow

HAZARD WARNING Copper sulphate is **toxic** if swallowed or if dust is inhaled.

EXPERIMENT 2.2 Determining an enthalpy change of solution

Since this experiment is designed as a planning exercise, students will probably be asked to submit their own list of requirements, but the following list should be adequate. The experiment is similar to Experiment 2.1.

Requirements per student (or pair)	Notes
■ weighing bottle	
■ teat pipette	students may ask for a burette
■ balance (nearest 0.01 g)	
■ polystyrene cup with lid	**expanded** polystyrene to minimise heat transfer, or two thin cups one inside the other. Lid needs hole for thermometer
■ safety spectacles	
■ spatula	
■ thermometer, 0–50 °C	must be graduated in 0.1 °C
■ ammonium chloride, NH_4Cl	about 5 g
■ distilled water	about 100 cm^3

EXPERIMENT 2.3 Using Hess' law (1)

Requirements per student (or pair)	Notes
■ 2 weighing bottles	
■ teat pipette	
■ balance (nearest 0.01 g)	
■ 2 polystyrene cups with lids	see Experiments 2.1 and 2.2
■ safety spectacles	
■ spatula	
■ thermometer, 0–50 °C	must be graduated in 0.1 °C
■ magnesium sulphate (anhydrous), $MgSO_4$	about 5 g. Heat to constant weight at about 250 °C and store in desiccator. If necessary, the hydrated salt may be dehydrated by prolonged heating (with stirring) to constant weight
■ magnesium sulphate-7-water, $MgSO_4{\cdot}7H_2O$	
■ distilled water	about 100 cm^3

EXPERIMENT 2.4 Determining enthalpies of combustion

Requirements per student (or pair)	Notes
■ 6 beakers, 50 cm^3	not required if alcohols are supplied in separate spirit lamps
■ 6 teat pipettes	
■ heat of combustion apparatus	
■ Drechsel bottle	to be fitted up as in diagram below. Consult teacher if alternative apparatus is used
■ filter pump	
■ rubber tubing for connections	
■ small wood block	
■ spirit lamp(s), small	7 if possible, two filled with propanol and one for each of the other alcohols
■ thermometer 0–50 °C (in 0.1 °C)	
■ balance (preferably to 0.001 g)	
■ Bunsen burner and protective mat	
■ labels	for the spirit lamp(s)
■ retort stand with 2 clamps and bosses	must be big enough to hold the heat of combustion apparatus
■ safety spectacles	
■ tweezers	**not** plastic-tipped – to adjust wick
■ wood splints	
■ tap water in large bottle	$2\frac{1}{2}$ dm^3. Stand in the laboratory overnight to get constant temperature
■ propan-1-ol, C_3H_7OH ■ butan-1-ol, C_4H_9OH ■ pentan-1-ol, $C_5H_{11}OH$ ■ hexan-1-ol, $C_6H_{13}OH$ ■ heptan-1-ol, $C_7H_{15}OH$ ■ octan-1-ol, $C_8H_{17}OH$	a few cm^3 of each. If possible they should be in labelled spirit lamps

HAZARD WARNING

The alcohols you will be using are highly flammable and are harmful by absorption through the skin and lungs. Therefore you **must**:

- **keep the stoppers on the bottles when not in use;**
- **keep the bottles away from flames;**
- **wash your hands after use (or wear gloves);**
- **wear safety spectacles.**

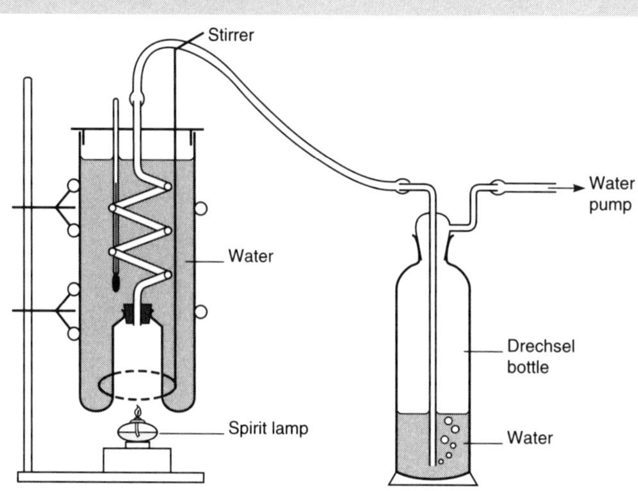

EXPERIMENT 2.5 A thermometric titration

Requirements per student (or pair)	Notes
■ burette, 50 cm^3, and stand	
■ filter funnel, small	to fill burette
■ pipette, 50 cm^3	25 cm^3 can be used
■ pipette filler	
■ polystyrene cup	
■ safety spectacles	
■ thermometer, 0–50 °C	graduated in 0.1 °C
■ ethanoic (acetic) acid, ~2.0 M CH_3CO_2H	about 50 cm^3. 115 cm^3 of glacial ethanoic acid made up to 1000 cm^3. Wear gloves and eye protection
■ hydrochloric acid, ~2.0 M HCl	about 50 cm^3. 172 cm^3 of concentrated HCl made up to 1000 cm^3. Wear gloves and eye protection
■ sodium hydroxide solution, 1.00 M NaOH	about 100 cm^3. If possible, make up precisely 1.00 M NaOH from a purchased ampoule. Otherwise dissolve 40.0 g of NaOH pellets in water, make up to 1000 cm^3, and standardise against a known acid
	All three solutions must be at the same temperature – make up the day before and store in the same room

HAZARD WARNING

Sodium hydroxide solution is corrosive. Ethanoic acid and hydrochloric acid solutions are irritants. Therefore you **must**:

- **use the pipette filler supplied;**
- **wear safety spectacles.**

EXPERIMENT 2.6 Using Hess' law (2)

Requirements per student (or pair)	Notes
■ 2 beakers, 100 cm^3	
■ pipette, 50 cm^3	25 cm^3 can be used
■ 2 weighing bottles	
■ balance (to 0.01 g)	
■ Bunsen burner, tripod, gauze and bench protection mat	
■ safety spectacles	
■ spatula	
■ thermometer, 0–100 °C	
■ vacuum flask with thermometer fitted (0–50 °C)	thermometer must be graduated in 0.1 °C. See below for fitting
■ anhydrous copper(II) sulphate, $CuSO_4$	about 5 g, preferably in desiccator. See below for preparation
■ copper(II) sulphate-5-water, $CuSO_4{\cdot}5H_2O$	about 7 g. Grind finely with pestle and mortar for quick dissolving
■ distilled water	about 350 cm^3. Tap water can be used for the first part **only** (250 cm^3)

HAZARD WARNING Copper sulphate is harmful is swallowed. Therefore you **must**:
- **use the pipette filler supplied;**
- **wear safety spectacles and gloves when making up the solutions.**

1. Insert a borer of size just larger than the thermometer into the hole in the bung from the narrow end.
2. Insert the thermometer to the required depth inside the borer.
3. Maintain the position of the thermometer while withdrawing the borer.
4. Remove the thermometer in a similar way after the experiment.

Anhydrous copper sulphate It is not as easy as it sounds to dehydrate the blue crystals completely without overheating which causes a little dark copper oxide to appear. It is best to purchase the anhydrous salt each year but if this is not practicable, proceed as follows:
1. Grind the required amount to powder, remembering that 25 g crystals reduces to 16 g. There is no point in dehydrating a lot more than will be used in a few days.
2. Put in a shallow dish (evaporating dish) of suitable size on a tripod and gauze. The powder should not be more than about 1 cm deep.
3. Heat over a moderate (not roaring) flame for at least 15 minutes, stirring constantly. Reduce heat if there is any sign of darkening in colour. Four molecules of water are lost very easily, but the fifth takes longer.
4. The only way to be certain that hydration is complete is to heat to constant weight.
5. Allow to cool somewhat before storing in a desiccator.

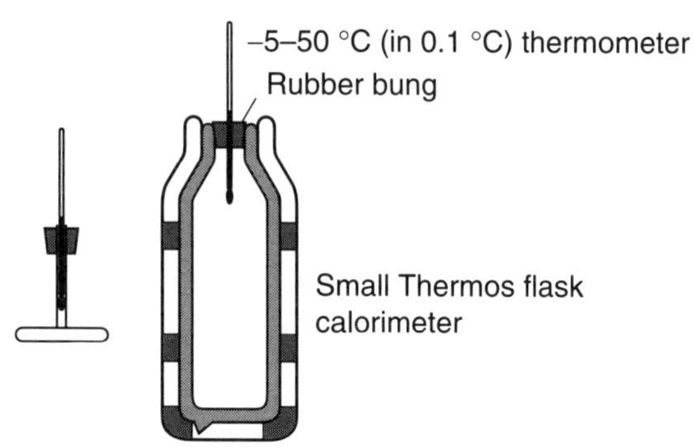

N.B. Copper sulphate is harmful if swallowed and solutions should not be poured away without much dilution. You may wish to keep the solutions returned by the students for use in junior work – they should be uncontaminated.

EXPERIMENT 3.1 Making models of two metallic structures

Requirements per student (or pair)	Notes
■ 31 expanded polystyrene spheres	in a bag or box! Any size will do as long as they are all the same but large spheres are easier to handle.
■ Blu-tak or similar demountable adhesive	Blu-tak is better for this purpose than Pritt 'Buddies'

EXPERIMENT 3.2 Recognising ionic, covalent and metallic structures

Requirements per student (or pair)	Notes
■ beaker, 100 cm^3	
■ 6 ignition tubes	
■ 6 test-tubes	
■ test-tube holder	to fit both tube sizes
■ battery, 6 V	
■ bulb, 6.5 V, 0.3 A	
■ bulb holder	
■ Bunsen burner and bench mat	
■ 2 carbon electrodes	
■ 3 connecting leads with crocodile clips	only 2 needed if bulb holder is permanently mounted on battery
■ safety spectacles	
■ barium sulphate, $BaSO_4$	A
■ copper(II) oxide, CuO	B
■ graphite, C, powdered	C
■ glucose, $C_6H_{12}O_6$	D
■ potassium bromide, KBr	E
■ zinc, Zn, powdered	F
■ silica, SiO_2, precipitated (or fine white sand)	G

A–G: a few grams of each in bottles or tubes labelled with **letter only**

EXPERIMENT 3.3 Testing liquids for polarity

Requirements per student (or pair)	Notes
■ 6 beakers, 250 cm^3	must be dry
■ 6 burettes in stands, each corked and labelled and containing one of the following liquids: cyclohexane, C_6H_{12} cyclohexene, C_6H_{10} distilled water, H_2O hexan-1-ol, $C_6H_{13}OH$ methyl ethanoate, $CH_3CO_2CH_3$ propanone, CH_3COCH_3	set up in the fume cupboard. Burettes should all be of the same type so that they give the same rate of flow, and the same volume of liquid (at least 25 cm^3) placed in each. The experiment may still be done if some of the liquids are not available. One set will serve a large class
■ 6 small funnels	for refilling burettes
■ rod of polythene or perspex or ebonite	must be quite dry
■ piece of fur or suitable cloth	must be quite dry. It is worth experimenting to find a good combination so that the rod is readily charged when rubbed with the cloth

HAZARD WARNING Cyclohexene has an irritant vapour.
All the organic liquids mentioned in the list are flammable.
Therefore you **must:**

- **perform this experiment in a fume cupboard;**
- **keep the stoppers on the bottles as much as possible;**
- **wear gloves and safety spectacles.**

Note that hexane, tetrachloromethane and trichloromethane should **not** now be used because of concerns about toxicity and/or ozone depletion.

Disposal of organic residues

Provided students have worked carefully the liquids may be returned to stock for re-use – perhaps in separate bottles, labelled accordingly. However, if liquids have been mixed or otherwise contaminated they may be disposed of as follows: **up to 10 cm^3 at a time** – emulsify with water and detergent and flush down the drain with **plenty** of water. **Larger volumes** – consult your local education authority or water authority.

EXPERIMENT 3.4 The effect of hydrogen bonding on liquid flow

Requirements per student (or pair)

- stopclock or stopwatch
- 4 long glass tubes, sealed at one end and securely corked at the other, each containing one of the following liquids:

  propan-1-ol, $CH_3CH_2CH_2OH$

  propan-1,2-diol, $CH_3CH(OH)CH_2OH$

  propan-1,2,3-triol (glycerol), $CH_2(OH)CH(OH)CH_2OH$

  propan-1,2,3-triyl triethanoate (glycerol triacetate), $CH_2(OCOCH_3)CH(OCOCH_3)CH_2OCOCH_3$

Notes

tubes must be labelled and of equal length (~1 m) and diameter (~8 mm). Each must have an air bubble of the same size (~20 mm). One set will serve a large class, and the tubes may be stored for other occasions. Corks must fit well enough to allow the tubes to be inverted

HAZARD WARNING

Propan-1-ol is highly flammable.
Propan-1,2-diol is harmful if swallowed.

EXPERIMENT 3.5 Variation of boiling point with composition of a mixture

Requirements per student (or pair)	Notes
■ 2 burettes, 50 cm^3, and stands	
■ 2 funnels, small	for filling burettes and flasks
■ ground-glass-joint apparatus	for **one** of the sets shown below
■ test-tube	
■ thermometer, 0–100 °C ± 0.2 °C	a long stem thermometer, graduated in 1°, will serve if necessary. A 50–100 °C thermometer, graduated in 0.1° would be ideal **provided** the scale is fully visible
■ anti-bumping granules	
■ Bunsen burner, gauze, tripod and mat	
■ safety spectacles	
■ cyclohexane, C_6H_{12} ■ ethanol, C_2H_5OH ■ propan-1-ol, $CH_3CH_2CH_2OH$ ■ propan-2-ol, $CH_3CH(OH)CH_3$	~20 cm^3 of each liquid. Each student will probably need only **two** of these – consult the teacher. IMS (74 op) will serve as ethanol. disposal of organic solvent **residues:** see Experiment 3.6

HAZARD WARNING Ethanol, cyclohexane, propan-1-ol and propan-2-ol are highly flammable.
Therefore you **must:**
- **keep the stoppers on the bottles as much as possible;**
- **keep the bottles away from flames;**
- **wear safety spectacles.**

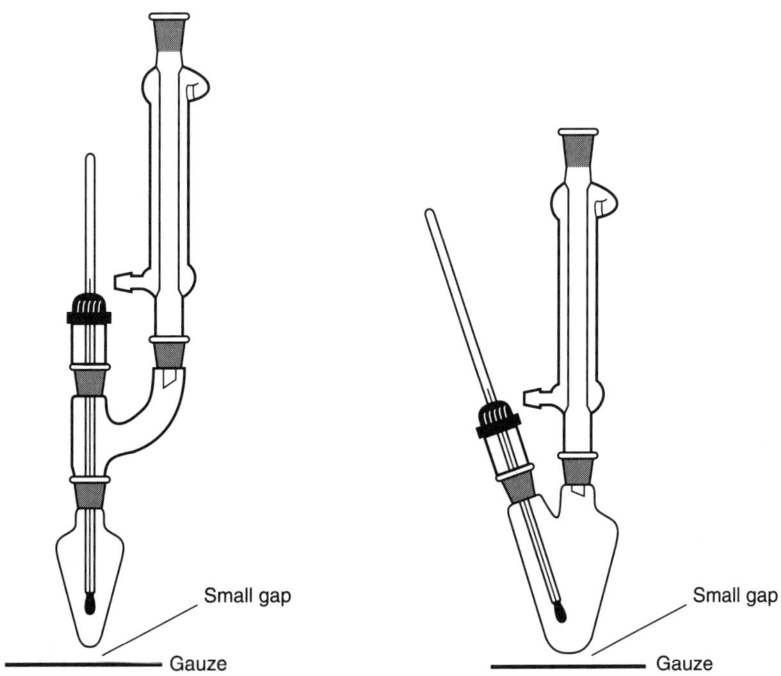

EXPERIMENT 3.6 Measuring the temperature change on forming solutions

Requirements per student (or pair)	Notes
■ beaker, 250 cm^3	preferably tall form, but not essential
■ boiling-tube	
■ 2 funnels, small	for transferring liquid
■ 2 measuring cylinders, 10 cm^3	
■ thermomcter, –5 to 50 °C ± 0.1 °C	
■ cotton wool	or similar insulating material, sufficient to fill beaker
■ safety spectacles	
■ cyclohexane, C_6H_{12}	~10 cm^3 of each. IMS (74 op) is suitable for ethanol. The experiment is simplified if all are kept at the same temperature for 24 hours beforehand
■ ethanol, C_2H_5OH	
■ propan-1-ol, $CH_3CH_2CH_2OH$	
■ propan-2-ol, $CH_3CH(OH)CH_3$	

HAZARD WARNING Ethanol, cyclohexane, propan-1-ol and propan-2-ol are highly flammable.
Therefore you **must:**

- **keep the stoppers on the bottles as much as possible;**
- **keep the bottles away from flames;**
- **wear safety spectacles.**

Disposal of organic solvent residues

The mixture of propanols may be simply washed down the drain with plenty of water.
The cyclohexane/ethanol mixture should be treated as follows: **up to 10 cm^3 at a time** – emulsify with water and detergent and flush down the drain with plenty of water. **Larger volumes** – consult your local education authority or water authority.

EXPERIMENT 4.1 Reaction between sodium peroxide and water

Requirements per student (or pair)	Notes
■ 3 test-tubes	
■ 3 teat pipettes	
■ Bunsen burner	
■ bench protection mat	
■ safety spectacles	
■ spatula	
■ test-tube rack	
■ wood splints (2 or 3)	
■ distilled water	
■ dilute sulphuric acid, 1 M H_2SO_4	2–3 cm^3. Carefully add 55 cm^3 of concentrated acid, slowly and with constant stirring, to 800 cm^3 of distilled water and make up to 1000 cm^3
■ 2-methylbutan-1-ol (*iso*-amyl alcohol), $C_5H_{11}OH$	2–3 cm^3. (Pentan-1-ol will serve)
■ potassium dichromate solution, 0.02 M $K_2Cr_2O_7$	2–3 cm^3. Dissolve 0.5 g of solid in distilled water and make up to 100 cm^3
■ sodium peroxide, Na_2O_2	about 1 g. Fresh (the solid changes from yellow to white on exposure to air)
■ universal indicator solution	

HAZARD WARNING

2-Methylbutan-1-ol is highly flammable. Therefore you **must**:

- **keep the stopper on the bottle when not in use;**
- **keep the liquid away from flames.**

Sodium peroxide is corrosive and a powerful oxidant. Therefore you **must:**

- **wear safety spectacles;**
- **avoid contact with skin.**

Concentrated sulphuric acid can cause severe burns and reacts vigorously with water. Therefore you **must**:

- **wear gloves and eye protection and follow instructions for dilution carefully.**

TECHNICIAN'S SHEET

EXPERIMENT 4.2 Heating the nitrates and carbonates of s-block elements

Requirements per student (or pair)	Notes
■ 30 test-tubes	
■ right-angled delivery tube, glass, with bung	see diagram below
■ 2 Bunsen burners	
■ 2 bench protection mats	
■ labels for test-tubes	
■ retort stand, clamp and boss	
■ safety spectacles	
■ 2 spatulas	
■ 2 stopclocks or stopwatches	
■ 2 test-tube holders	
■ 2 test-tube racks	student needs **four** rows of 4 or 5 tubes each
■ wood splints	
■ nitrates and anhydrous carbonates of lithium, sodium, potassium, rubidium, caesium, magnesium, calcium, strontium and barium.	a few grams of each available compound in the form of powder or fine crystals
■ hydrochloric acid, dilute, 2 M HCl	about 10 cm^3. Add 172 cm^3 of concentrated acid to 800 cm^3 of distilled water and make up to 1000 cm^3

HAZARD WARNING

Barium compounds are **toxic** if swallowed or absorbed through the skin.
■ **Avoid contact with skin.**
Nitrogen dioxide is a toxic gas.
■ **Heat nitrates in a fume cupboard.**
Nitrates are strong oxidising agents.
■ **Do not allow pieces of glowing splint to drop onto hot nitrates.**
Concentrated hydrochloric acid is corrosive.
■ **Wear eye protection and gloves when diluting.**

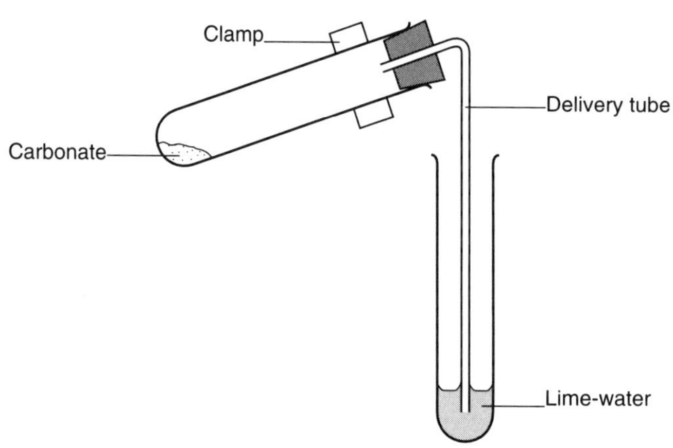

EXPERIMENT 4.3 The solubility of some salts of Group II elements

Requirements per student (or pair)	Notes
■ 16 test-tubes	
■ 5 teat pipettes marked at 1 cm^3	draw in 1 cm^3 of water from a 10 cm^3 measuring cylinder and mark with a small blob of paint
■ labels for test-tubes	
■ safety spectacles	
■ 4 test-tube racks	
■ distilled water	
■ 8 solutions labelled as follows:	a few cm^3 of each
0.1 M $Ba^{2+}(aq)$	26 g $Ba(NO_3)_2$ or 24 g $BaCl_2{\cdot}2H_2O$ per dm^3
0.1 M $Ca^{2+}(aq)$	24 g $Ca(NO_3)_2{\cdot}4H_2O$ or 11g $CaCl_2$ per dm^3
0.1 M $Mg^{2+}(aq)$	26 g $Mg(NO_3)_2{\cdot}6H_2O$ or 20 g $MgCl_2{\cdot}6H_2O$ per dm^3
0.1 M $Sr^{2+}(aq)$	21 g $Sr(NO_3)_2$ or 27 g $SrCl_2{\cdot}6H_2O$ per dm^3
0.05 M $CO_3^{2-}(aq)$	5.3 g Na_2CO_3 (anhydrous) per dm^3
1.0 M $OH^-(aq)$	40 g NaOH per dm^3 (freshly made)
0.5 M $SO_3^{2-}(aq)$	53 g Na_2SO_3 or 126 g $Na_2SO_3{\cdot}7H_2O$ per dm^3
0.5 M $SO_4^{2-}(aq)$	71 g Na_2SO_4 or 161 g $Na_2SO_4{\cdot}10H_2O$ per dm^3

HAZARD WARNING

1.0 M solutions of OH^- ions are corrosive. Therefore you **must**:

- ■ **wear safety spectacles, especially when preparing solutions.**

Solutions of sulphite ions, SO_3^{2-}, release a **toxic** gas, sulphur dioxide if acid is added. Therefore you **must**:

- ■ **take care with disposal of these solutions, especially if you suffer from asthma.**

EXPERIMENT 4.4 The solubility of the halogens in organic solvents

Requirements per student (or pair)	Notes
■ 3 test-tubes, each with a bung or cork	
■ 5 teat pipettes, marked at 1 cm^3	draw in 1 cm^3 of water from a 10 cm^3 measuring cylinder and mark with a spot of paint
■ safety spectacles	
■ test-tube rack	
■ 3 reagent bottles for organic residues	clearly labelled 'cyclohexane residues', 'Volasil residues' and 'ethoxyethane residues'. See below for disposal
■ bromine water	about 5 cm^3. Carefully break a 1 cm^3 phial under 100 cm^3 of water and shake to dissolve
■ chlorine water	about 5 cm^3. Generate chlorine by adding concentrated hydrochloric acid dropwise to solid potassium manganate(VII) (permanganate) and bubble through water in a fume cupboard
■ iodine solution, 0.05 M I_2 in KI(aq)	about 5 cm^3. Dissolve 2 g of potassium iodide, KI, in 20 cm^3 of distilled water, add 1.25 g of iodine and stir to dissolve. Dilute to 100 cm^3
■ ethoxyethane (diethyl ether), $(C_2H_5)_2O$	about 10 cm^3
■ cyclohexane, C_6H_{12}	about 10 cm^3
■ Volasil 244*	about 10 cm^3

HAZARD WARNING

- This experiment should be performed in a **fume cupboard** since all the halogen vapours are **toxic** and the organic vapours are dangerous to inhale. If a fume cupboard is not available then **the laboratory must be well ventilated.**
- Check that no flames are close to where you are working because ethoxyethane, cyclohexane and Volasil are extremely flammable.

Disposal of organic residues

Small quantities (i.e. those generated by experiments such as this in the majority of schools) may be poured down a sink in a fume cupboard with plenty of flowing water and the addition of some detergent. For disposal of larger quantities, consult your local education authority or water authority. Normally, residues should not be stored, nor should residues from different experiments, containing different chemicals, be mixed.

*Volasil 244 is a cyclic organosilicone, octamethylcyclotetrasiloxane. Its molecule has alternating silicon and oxygen atoms in a ring structure with methyl groups attached to the silicon atoms. We have included it here, and subsequently, because it is more environmentally friendly than most organic solvents and is safer to use, although it is flammable. Volasil 244 may be obtained from BDH Laboratory Supplies, Merck Ltd, Hunter Boulevard, Lutterworth, LE17 4XN. Tel: 0800 223344. (1997 prices: £12.80 for 500 cm^3, £27.10 for 2.5 litres. Catalogue number 63300.)

EXPERIMENT 4.5 The action of dilute alkali on the halogens

Requirements per student (or pair)	Notes
■ 5 test-tubes	
■ 4 teat pipettes	
■ safety spectacles	
■ test-tube rack	
■ wash-bottle of distilled water	
■ bromine water	about 2 cm^3. Carefully break a 1 cm^3 phial under 100 cm^3 of distilled water and shake to dissolve
■ iodine solution, 0.05 M I_2 in KI(aq)	about 2 cm^3. Dissolve 2 g of potassium iodide, KI, in 20 cm^3 of distilled water, add 1.25 g of iodine and stir to dissolve. Dilute to 100 cm^3 with water
■ sodium hydroxide solution, 2 M NaOH	dissolve 80 g of sodium hydroxide pellets in distilled water and make up to 1000 cm^3
■ sulphuric acid, dilute, 1 M H_2SO_4	carefully add 55 cm^3 of concentrated suphuric acid slowly and with constant stirring, to 800 cm^3 of distilled water and make up to 1000 cm^3

HAZARD WARNING

Bromine water is **toxic** and corrosive. The vapour is extremely irritant to the eyes, lungs and skin. Therefore you **must**:

■ **avoid contact with skin;**

■ **avoid inhaling the vapour.**

Concentrated sulphuric acid can cause severe burns and reacts vigorously with water. Therefore you **must**:

■ **wear gloves and eye protection and follow instructions for dilution carefully.**

Sodium hydroxide is corrosive:

■ **wear gloves and eye protection when making up solutions.**

TECHNICIAN'S SHEET

EXPERIMENT 4.6 Halogen–halide reactions in aqueous solution

Requirements per student (or pair)	Notes
■ 6 test-tubes, each with a bung or cork	
■ 7 teat pipettes	
■ safety spectacles	
■ reagent bottle for organic residues	labelled 'Volasil 244 residues' or 'cyclohexane residues'
■ test-tube rack	
■ bromine water	about 5 cm^3. Carefully break a 1 cm^3 phial under 100 cm^3 of distilled water and shake to dissolve
■ chlorine water	about 5 cm^3. Generate chlorine by adding concentrated hydrochloric acid dropwise to solid potassium manganate(VII) (permanganate) and bubble through water in a fume cupboard
■ Volasil 244* or cyclohexane, C_6H_{12}	about 10 cm^3
■ iodine solution, 0.05 M I_2 in KI(aq)	about 5 cm^3. Dissolve 2 g of potassium iodide, KI, in 20 cm^3 of distilled water, add 1.25 g of iodine and stir to dissolve. Dilute to 100 cm^3
■ potassium bromide solution, 0.1 M KBr	about 5 cm^3. Dissolve 1.2 g of solid in distilled water and make up to 100 cm^3
■ potassium chloride solution, 0.1 M KCl	about 5 cm^3. Dissolve 0.75 g of solid in distilled water and make up to 100 cm^3
■ potassium iodide solution, 0.1 M KI	about 5 cm^3. Dissolve 1.7 g of solid in distilled water and make up to 100 cm^3

HAZARD WARNING

Halogen vapours must not be inhaled. If a fume cupboard is not available then:
- **the laboratory must be well ventilated and reagent bottles and test-tubes stoppered as much as possible;**
- **keep Volasil 244 away from flames.**

Concentrated hydrochloric acid is corrosive. Therefore you **must**:
- **wear eye protection and gloves when diluting.**

*Volasil 244 is a cyclic organosilicone, octamethylcyclotetrasiloxane. Its molecule has alternating silicon and oxygen atoms in a ring structure with methyl groups attached to the silicon atoms. We have included it here, and subsequently, because it is more environmentally friendly than most organic solvents and is safer to use, although it is flammable. Volasil 244 may be obtained from BDH Laboratory Supplies, Merck Ltd, Hunter Boulevard, Lutterworth, LE17 4XN. Tel: 0800 223344. (1997 prices: £12.80 for 500 cm^3, £27.10 for 2.5 litres. Catalogue number 63300.)

EXPERIMENT 4.7 Reactions of solid halides

Requirements per student (or pair)	Notes
■ 1 stirring rod, glass	preferably hard glass
■ 12 test-tubes, and test-tube rack	
■ 3 teat pipettes	
■ Bunsen burner and protective mat	
■ filter paper strips	for making test papers
■ safety spectacles	
■ spatula	
■ test-tube holder	for heating. Metal type preferred
■ wash-bottle of distilled water	about 2 cm^3, preferably in a dropping bottle
■ ammonia solution, 0.880 NH_3	about 5 cm^3
■ Volasil 244* or cyclohexane, C_6H_{12}	about 5 cm^3. Dissolve 3.3 g of solid trihydrate in distilled water and make up to 100 cm^3
■ lead(II) ethanoate (acetate) solution, 0.1 M $(CH_3CO_2)_2Pb$	
■ manganese(IV) oxide, MnO_2	about 2 g
■ potassium bromide, KBr	about 2 g
■ potassium chloride, KCl	about 2 g
■ potassium iodide, KI	about 2 g
■ phosphoric(V) acid, 100%, H_3PO_4 (orthophosphoric acid)	about 1 g. If the solid is not available, cautiously add a little phosphorus(V) oxide, P_4O_{10}, to 'syrupy' phosphoric acid (90%)
■ potassium dichromate(VI) solution, 0.1 M $K_2Cr_2O_7$	about 5 cm^3. Dissolve 2.9 g in distilled water and make up to 100 cm^3
■ starch solution, 1%	about 5 cm^3. Mix 1 g of **soluble** starch to a paste with a little water. Add the paste to 100 cm^3 of boiling distilled water, with stirring, and boil for 1 minute. Cool. (Keep no longer than two weeks)
■ sulphuric acid, concentrated, H_2SO_4	about 2 cm^3

HAZARD WARNING

Concentrated sulphuric acid is very corrosive and reacts violently with water. Phosphoric(V) acid and potassium dichromate are also very corrosive. Lead(II) ethanoate is **toxic**. Therefore, you **must**:

- **avoid contact with skin; if contact does occur, wash immediately under a cold tap with plenty of water;**
- **dispose of cold residues containing concentrated sulphuric acid by pouring slowly into plenty of water.**

The halogens, hydrogen halides, sulphur dioxide and hydrogen sulphide are **toxic**.

- **Carry out these experiments in a fume cupboard.**

Volasil 244 and cyclohexane are very flammable.

- **Keep Volasil and cyclohexane stoppered and away from flames.**

*Volasil 244 is a cyclic organosilicone, octamethylcyclotetrasiloxane. Its molecule has alternating silicon and oxygen atoms in a ring structure with methyl groups attached to the silicon atoms. We have included it here, and subsequently, because it is more environmentally friendly than most organic solvents and is safer to use, although it is flammable. Volasil 244 may be obtained from BDH Laboratory Supplies, Merck Ltd, Hunter Boulevard, Lutterworth, LE17 4XN. Tel: 0800 223344. (1997 prices: £12.80 for 500 cm^3, £27.10 for 2.5 litres. Catalogue number 63300.)

EXPERIMENT 4.8 Reactions of halides in solution

Requirements per student (or pair)	Notes
■ 18 test-tubes, 9 with corks or bungs, and 3 racks	
■ 8 teat pipettes	
■ safety spectacles	
■ ammonia solution, 5 M NH_3	about 15 cm^3. Dilute 28 cm^3 of 0.880 ammonia to 100 cm^3 with distilled water
■ Volasil 244* or cyclohexane, C_6H_{12}	about 5 cm^3
■ hydrogen peroxide solution, H_2O_2	about 5 cm^3. Dilute 30 cm^3 of '10-volume' hydrogen peroxide to 100 cm^3 with distilled water
■ lead(II) nitrate solution, 0.1 M $Pb(NO_3)_2$	about 5 cm^3. Dissolve 1.7 g in distilled water and make up to 100 cm^3
■ nitric acid, dilute, 2 M HNO_3	about 15 cm^3. Carefully add 120 cm^3 of concentrated nitric acid to about 700 cm^3 of distilled water and make up to 1000 cm^3
■ potassium bromide solution, 0.1 M KBr	about 5 cm^3. Dissolve 1.2 g of solid in distilled water and make up to 100 cm^3
■ potassium chloride solution, 0.1 M KCl	about 5 cm^3. Dissolve 0.75 g of solid in distilled water and make up to 100 cm^3
■ potassium iodide solution, 0.1 M KI	about 5 cm^3. Dissolve 1.7 g of solid in distilled water and make up to 100 cm^3
■ silver nitrate solution, 0.02 M $AgNO_3$	about 10 cm^3. Dissolve 0.30 g of solid in distilled water and make up to 100 cm^3
■ starch solution 1%	about 5 cm^3. Mix 1 g of **soluble** starch to a paste with a little water. Add the paste to 100 cm^3 of boiling distilled water, with stirring, and boil for 1 minute. Cool. (Keep for no longer than two weeks)
■ sulphuric acid, dilute, 1 M H_2SO_4	about 10 cm^3. Carefully add 55 cm^3 of concentrated sulphuric acid, slowly and with constant stirring, to 800 cm^3 of distilled water and make up to 1000 cm^3

HAZARD WARNING

■ **Keep Volasil 244 well stoppered and away from flames.**

Silver nitrate is **toxic** and corrosive. Lead nitrate is **toxic.**

Ammonia solution and hydrogen peroxide are corrosive substances.

Concentrated sulphuric acid can cause severe burns and reacts vigorously with water. Therefore you **must**:

■ **wear gloves and eye protection and follow instructions for dilution carefully.**

Concentrated nitric acid is very corrosive and a powerful oxidant. Therefore you **must**:

■ **wear gloves and eye protection when diluting.**

*Volasil 244 is a cyclic organosilicone, octamethylcyclotetrasiloxane. Its molecule has alternating silicon and oxygen atoms in a ring structure with methyl groups attached to the silicon atoms. We have included it here, and subsequently, because it is more environmentally friendly than most organic solvents and is safer to use, although it is flammable. Volasil 244 may be obtained from BDH Laboratory Supplies, Merck Ltd, Hunter Boulevard, Lutterworth, LE17 4XN. Tel: 0800 223344. (1997 prices: £12.80 for 500 cm^3, £27.10 for 2.5 litres. Catalogue number 63300.)

EXPERIMENT 4.9 Balancing a redox reaction

Requirements per student (or pair)	Notes
■ 2 burettes, 50 cm^3	
■ 2 conical flasks, 250 cm^3	
■ 2 funnels, small	for filling burettes
■ 2 measuring cylinders, 10 cm^3	
■ 2 burette stands	
■ safety spectacles	
■ wash-bottle of distilled water	
■ white tile	
■ hydrochloric acid, 2 M HCl	about 50 cm^3. Dilute 172 cm^3 of concentrated hydrochloric acid to 1000 cm^3 with distilled water
■ potassium iodate(V) solution, 0.10 M KIO_3	about 25 cm^3. Dissolve 21.40 g of AR KIO_3 in distilled water and make up to 1000 cm^3 in a volumetric flask
■ potassium iodide solution, 1 M KI	about 50 cm^3. Dissolve 166 g of solid in distilled water and make up to 1000 cm^3
■ sodium thiosulphate solution, 0.10 M $Na_2S_2O_3$	about 100cm^3. Dissolve 24.8 g of AR $Na_2S_2O_3{\cdot}5H_2O$ in distilled water and make up to 1000 cm^3 in a volumetric flask
■ starch solution, 0.2%	about 10 cm^3. Mix 0.2 g of **soluble** starch to a paste with a little distilled water. Add the paste to 100 cm^3 of boiling water, with stirring, and boil for 1 minute. Cool. (Keep for no longer than a day or two)

HAZARD WARNING

Concentrated hydrochloric acid is corrosive. Therefore you **must**:

■ **wear eye protection and gloves when diluting.**

EXPERIMENT 4.10 Observation and deduction exercise 1

Requirements per student (or pair)	Notes
■ 5 test-tubes	preferably hard glass
■ Bunsen burner and protective mat	
■ filter paper strips	for making test papers
■ litmus papers, blue and red	
■ safety spectacles	
■ spatula	
■ test-tube holder	
■ test-tube rack	
■ wash-bottle of distilled water	
■ wood splints	
■ potassium bromate(V), $KBrO_3$	about 1 g. **Unknown** – labelled 'F'
■ chlorine water	about 2 cm^3. Generate chlorine by adding concentrated hydrochloric acid dropwise to solid potassium manganate(VII) (permanganate) and bubble through water **in a fume cupboard**
■ lead(II) ethanoate (acetate) solution, 0.1 M $(CH_3CO_2)_2Pb$	about 2 cm^3. Dissolve 3.3 g of solid trihydrate in distilled water and make up to 100 cm^3
■ nitric acid, dilute, 2 M HNO_3	about 5 cm^3. Add 128 cm^3 of concentrated acid slowly to 800 cm^3 of distilled water and make up to 1000 cm^3
■ silver nitrate solution, 0.02 M $AgNO_3$	about 1 cm^3. Dissolve 0.30 g of solid in distilled water and make up to 100 cm^3
■ students may also ask for the following (issued only on request): lime-water, $Ca(OH)_2$ (aq)	about 2 cm^3 of saturated solution

HAZARD WARNING

Silver nitrate is **toxic** and corrosive. Lead ethanoate is **toxic**.
Concentrated nitric acid is very corrosive and a powerful oxidant. Therefore you **must**:

- **wear gloves and eye protection when diluting.**

Chlorine gas is **toxic**. Therefore you **must**:

- **control the flow of gas with care when making a solution.**

THE HALOGENS

EXPERIMENT 4.11 Observation and deduction exercise 2

Requirements per student (or pair)	Notes
■ 5 test-tubes, and test-tube rack	preferably hard glass test-tubes
■ Bunsen burner and protective mat	
■ filter paper strips	for making test papers
■ litmus papers, red and blue	
■ safety spectacles	
■ spatula	
■ test-tube holder	preferably metal
■ wash-bottle of distilled water	
■ wood splints	
■ potassium iodide, KI	about 2 g. **Unknown** – labelled 'D'
■ potassium iodate(V), KIO_3	about 2 g. **Unknown** – labelled 'E'
■ sulphuric acid, concentrated, H_2SO_4	about 1 cm^3
■ potassium manganate(VII) solution, 0.01 M $KMnO_4$ (permanganate)	about 5 cm^3. Dissolve 1.6 g of solid in water and make up to 1000 cm^3
■ sulphuric acid, dilute, 1 M H_2SO_4	about 10 cm^3. Carefully add 55 cm^3 of concentrated sulphuric acid, slowly and with constant stirring, to 800 cm^3 of distilled water and make up to 1000 cm^3
■ students **may** also ask for the following (issued only on request):	
ammonia solution, 0.880 NH_3	about 2 cm^3
lime-water, $Ca(OH)_2$ (aq)	about 2 cm^3 of saturated solution
Volasil 244* or cyclohexane, C_6H_{12}	about 2 cm^3
nitric acid, dilute, 2 M HNO_3	about 5 cm^3. See Experiment 4.10
silver nitrate solution, 0.02 M $AgNO_3$	about 1 cm^3. Dissolve 0.30 g of solid in distilled water and make up to 100 cm^3
lead(II) ethanoate solution, 0.1 M $(CH_3CO_2)_2Pb$	about 2 cm^3. Dissolve 3.3 g of solid trihydrate in distilled water and make up to 100 cm^3

HAZARD WARNING

Concentrated sulphuric acid is very corrosive and reacts violently with water. Therefore you **must**:

- **avoid contact with skin; if contact does occur, wash immediately under a cold tap with plenty of water;**
- **dispose of cold residues by pouring slowly into plenty of water;**
- **wear eye protection and gloves when diluting.**

Volasil 244 and cyclohexane are flammable. Therefore you **must**:

- **keep away from flames and replace the stopper.**

*Volasil 244 is a cyclic organosilicone, octamethylcyclotetrasiloxane. Its molecule has alternating silicon and oxygen atoms in a ring structure with methyl groups attached to the silicon atoms. We have included it here, and subsequently, because it is more environmentally friendly than most organic solvents and is safer to use, although it is flammable. Volasil 244 may be obtained from BDH Laboratory Supplies, Merck Ltd, Hunter Boulevard, Lutterworth, LE17 4XN. Tel: 0800 223344. (1997 prices: £12.80 for 500 cm^3, £27.10 for 2.5 litres. Catalogue number 63300.)

EXPERIMENT 4.12 Investigating the properties of Period 3 chlorides

Requirements per student (or pair)	Notes
■ 2 measuring cylinders, 10 cm^3	1 must be dry
■ 4 teat pipettes	
■ 14 test-tubes	6 must be dry
■ access to fume cupboard	
■ access to organic residues bottle	labelled 'cyclohexane residues'. See Experiment 4.4 for disposal instructions
■ gloves	disposable type is adequate
■ pH paper	to cover range 1 to 7
■ safety spectacles	
■ spatula	
■ test-tube rack	
■ thermometer, 0–100 °C (±1 °C)	
■ wash-bottle of distilled water	
■ aluminium chloride, $AlCl_3$	about 2 g, preferably in the anhydrous form
■ disulphur dichloride, S_2Cl_2	about 2 cm^3
■ cyclohexane, C_6H_{12}	about 35 cm^3
■ magnesium chloride, $MgCl_2$	about 2 g
■ phosphorus trichloride, PCl_3	about 2 cm^3
■ silicon tetrachloride, $SiCl_4$	about 2 cm^3. See below
■ sodium chloride, NaCl	about 2 g
■ ammonia solution, 2 M NH_3	about 2 cm^3
■ universal indicator solution	about 5 cm^3. Colour chart also needed

HAZARD WARNING

Many of the chlorides in this experiment react vigorously with water. $SiCl_4$, PCl_3 and S_2Cl_2 are corrosive and harmful. Therefore you **must:**

- **do this experiment in a fume cupboard;**
- **keep stoppers on bottles as much as possible;**
- **wear gloves and safety spectacles.**

Silicon tetrachloride is readily hydrolysed and the product can firmly seal a stopper in place and allow pressure of hydrogen chloride gas to build up in the bottle.
Bottles containing silicon tetrachloride have been known to burst in storage or on being opened. Bottles of $SiCl_4$ should be opened in the fume cupboard with considerable caution after covering the bottle with a cloth. It is particularly dangerous to return unused chemical to the bottle.
Cyclohexane is very flammable. Therefore you **must:**

- **keep the stopper on the bottle as much as possible;**
- **keep the bottle away from flames.**

EXPERIMENT 4.13 Preparing anhydrous aluminium chloride

Requirements per student (or pair)	Notes
■ absorption tube	straight form with one bulb. See diagram below for apparatus to be assembled by students
■ access to desiccator	
■ adaptor with 'T' connection	with ground-glass-joints
■ combustion tube	1 cm diameter or larger
■ cylindrical dropping funnel, 50 cm^3	containing ground-glass-joint and tap
■ glass rod	
■ pear-shaped flask, 50 cm^3	with ground-glass-joint
■ receiver bottle with bung	bung should be fitted with a short length of glass tubing to connect with the absorption tube. It should also have a hole to fit the combustion tube
■ short length of glass tubing fitted into bung	to make connection between combustion tube and adaptor fitted with lid
■ specimen (sample) tube	
■ access to balance capable of weighing to 0.01 g	
■ access to fume cupboard	
■ 3 bosses, clamps and stands	
■ Bunsen burner and bench mat	
■ ceramic wool	
■ forceps	
■ long spatula	
■ rubber tubing for connections	
■ ruler, 30 cm	
■ safety spectacles and disposable gloves	
■ sticky labels	
■ aluminium foil, Al	about 0.25 g
■ anhydrous calcium chloride, $CaCl_2$	fused granular 3-8 mesh. About 20 g
■ potassium manganate(VII) (permanganate), $KMnO_4$	about 5 g
■ hydrochloric acid, concentrated, HCl	about 10 cm^3
■ soda lime	about 5 g

HAZARD WARNING Chlorine is a **toxic** gas. Concentrated hydrochloric acid is a corrosive liquid, and its vapour is harmful to eyes, lungs and skin. Therefore you **must**:

- **do the experiment at the fume cupboard.**
- **wear gloves and safety spectacles;**
- **keep the hydrochloric acid in a fume cupboard;**
- **keep the stopper on the bottle as much as possible.**

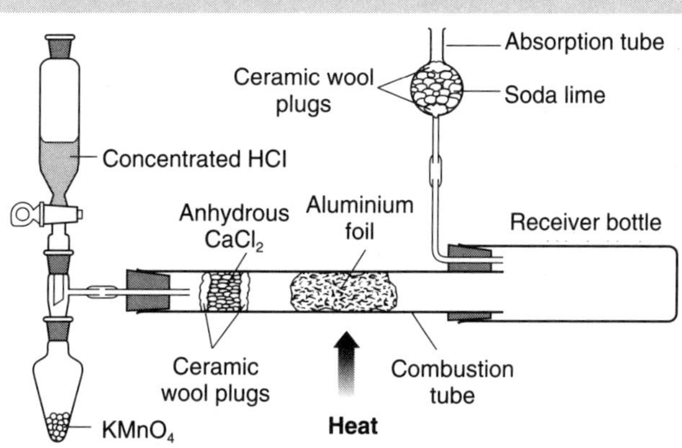

EXPERIMENT 4.14 Investigating the properties of Period 3 oxides

Requirements per student (or pair)	Notes
■ Drechsel bottle	fitted to SO_2 cylinder
■ measuring cylinder, 10 cm^3	
■ measuring cylinder, 100 cm^3	for bubbling SO_2 through water
■ teat pipette	
■ 6 test-tubes	
■ access to fume cupboard	
■ glass tubing	with right angled bend. One side should be long enough to reach the bottom of the 100 cm^3 measuring cylinder
■ pH paper	
■ rubber tubing for connections	
■ safety spectacles	
■ spatula	
■ splints	
■ test-tube rack	
■ thermometer, 0–100 °C ± 1 °C	
■ wash-bottle of distilled water	
■ aluminium oxide, Al_2O_3	about 2 g
■ magnesium oxide, MgO	about 2 g
■ phosphorus(V) oxide, P_4O_{10}	about 2 g
■ silicon(IV) oxide, SiO_2	about 2 g
■ sodium peroxide, Na_2O_2	about 2 g
■ sulphur dioxide cylinder, fitter with a delivery tube	or some other means of generating SO_2 gas
■ universal indicator solution	about 5 cm^3. Colour chart also needed

HAZARD WARNING

Phosphorus(V) oxide is corrosive and irritates eyes, skin and lungs. Sodium peroxide is also corrosive and a powerful oxidant. Sulphur dioxide is a toxic gas with a choking smell. Therefore you **must**:

- **do the experiment at the fume cupboard;**
- **wear safety spectacles;**
- **avoid contact with skin.**

ILPAC 5

INTRODUCTION TO ORGANIC CHEMISTRY

TECHNICIAN'S SHEET

EXPERIMENT 5.1 Chemical properties of alkanes

Requirements per student (or pair)	Notes
■ 5 test-tubes, dry, with corks to fit	
■ 1 test-tube, dry, covered with foil	aluminium foil should be tightly wrapped round the tube to prevent light from entering
■ watch glass, hard glass	
■ access to fume cupboard	
■ Bunsen burner and bench mat	
■ gloves	(disposable type is adequate)
■ safety spectacles	
■ lamp with 100 watt bulb	room fitting or table lamp
■ wood splints	
■ ammonia solution, 5 M	small bottle with glass stopper. See Experiment 5.2
■ bromine dissolved in Volasil 244 (with teat pipette)	about 2 cm^3. See Experiment 5.2
■ cyclohexane (with teat pipette)	about 6 cm^3
■ potassium manganate(VII) (permanganate) solution, 0.01 M $KMnO_4$	about 2 cm^3. Dissolve about 1 g in distilled water and make up to 1000 cm^3.
■ sulphuric acid, concentrated	about 2 cm^3. Small bottle with dropper placed on a drip tray
■ sulphuric acid, dilute, 1 M H_2SO_4	about 2 cm^3. See Experiment 5.2

HAZARD WARNING

Bromine is dangerously **toxic** and corrosive, especially in its liquid state. Solutions, such as those used in this experiment, must also be treated with care. Therefore you **must**:

- **do the experiment in a fume cupboard;**
- **keep the top on the bottle as much as possible;**
- **wear gloves and safety spectacles.**

Cyclohexane is very flammable. Therefore you **must**:

- **keep the top on the bottle as much as possible;**
- **keep the bottle away from flames;**
- **wear safety spectacles.**

Ammonia solution (5 M) is an irritant. Take great care when opening bottles on a hot day. You **must**:

- **wear safety spectacles.**

Concentrated sulphuric acid is very corrosive and reacts violently with water. Therefore you **must**:

- **wear gloves and safety spectacles;**
- **dispose of unwanted acid by cooling and pouring slowly into an excess of water.**

Note: Instructions for the preparations of the solutions are on the Technician's Sheet for Experiment 5.2. For disposal of residues see page 129.

EXPERIMENT 5.2 Chemical properties of alkenes

Requirements per student (or pair)

The requirements are exactly the same as for Experiment 5.1 except that **cyclohexene** is used instead of cyclohexane.

HAZARD WARNING Cyclohexene is an irritant. The vapour must not be inhaled. In the reaction between cyclohexene and concentrated sulphuric acid the amounts suggested must be strictly adhered to. Eye protection must be worn as usual and the front of the fume cupboard should be down as far as possible.
The reaction of ethene with bromine should never be carried out. The product 1,2-dibromoethane is a carcinogen.
All other precautions taken for Experiment 5.1 must also be adhered to here.

Preparation of solutions

Bromine dissolved in Volasil 244*

This should be prepared in a fume cupboard and protective gloves worn. Use 1 cm^3 of bromine in 125 cm^3 of Volasil 244. It is probably safest to crush an ampoule using a pair of pliers but care must be taken when cleaning out the vessel after use. Alternatively make a scratch on the neck of the ampoule with a good glass knife and snap off the neck. Label the bottle to warn of broken glass. This solution should be kept in an amber bottle in a ventilated store. It is stable over a few weeks but not months. After a few days, some gas may build up over the solution so, before handing it over to the students, release the stopper in a fume cupboard.
The reaction of ethene with bromine should never be carried out. The product 1,2-dibromoethane is a carcinogen.

Sulphuric acid, dilute 1 M H_2SO_4

Usual stock acid is very corrosive. Gloves and eye protection should be worn. Always add the acid to water and do so with care.

To prepare 1 litre of 1 M solution measure out 54 cm^3 of concentrated acid (98% w/v) in a measuring cylinder and put 600 cm^3 of distilled water in a 2 litre beaker. Pour the acid into the water in a thin stream while stirring with a glass rod using the other hand. Pour this solution into a litre measuring cylinder and make up to the mark with distilled water.

Ammonia solution, 5 M

To make 500 cm^3 measure 140 cm^3 of concentrated ammonia and add about 300 cm^3 of distilled water, stir, dilute to the final volume and stir again. Take great care opening bottles of 880 ammonia, particularly on hot days. A **face-shield** is recommended.

*Volasil 244 is a cyclic organosilicone, octamethylcyclotetrasiloxane. Its molecule has alternating silicon and oxygen atoms in a ring structure with methyl groups attached to the silicon atoms. We have included it here, and subsequently, because it is more environmentally friendly than most organic solvents and is safer to use, although it is flammable. Volasil 244 may be obtained from BDH Laboratory Supplies, Merck Ltd, Hunter Boulevard, Lutterworth, LE17 4XN. Tel: 0800 223344. (1997 prices: £12.80 for 500 cm^3, £27.10 for 2.5 litres. Catalogue number 63300.)

EXPERIMENT 5.3 Hydrolysis of organic halogen compounds

Requirements per student (or pair)	Notes
■ beaker, 250 cm^3	
■ measuring cylinder, 10 cm^3	
■ thermometer, 0–100 °C	
■ 5 test-tubes fitted with corks	
■ Bunsen burner, tripod, gauze and bench protection mat	the teacher may ask instead for a thermostatically controlled water-bath
■ safety spectacles	
■ clock with second hand	room clock or wrist-watches are adequate
■ protective gloves	disposable type preferred
■ test-tube rack	
■ waterproof marker or chinagraph pencil	
■ 1-bromobutane, C_4H_9Br	} ~1 cm^3 of each. Teat pipettes also required – see Introduction to Technician's Sheets (page 127)
■ 1-chlorobutane, C_4H_9Cl	}
■ 1-iodobutane, C_4H_9I	}
■ chlorobenzene, C_6H_5Cl	}
■ ethanol, C_2H_5OH	~20 cm^3. Industrial methylated spirit (IMS) is suitable
■ silver nitrate solution, 0.05 M $AgNO_3$	~5 cm^3. Dissolve 2.1 g of solid in distilled water and make up to 250 cm^3. Store and dispense in amber bottles

HAZARD WARNING

All organic halogen compounds have harmful vapours and can be **toxic** by absorption through the skin. Some are flammable. Therefore you **must**:

- **keep the stoppers on the bottles as much as possible;**
- **keep the bottles away from flames;**
- **wear safety spectacles and gloves.**

EXPERIMENT 5.4 Preparation of a halogenoalkane

Requirements per student (or pair)	Notes
■ conical flask, 100 cm^3 with bung	
■ ground-glass-joint apparatus shown in diagram below	14/23 joints and 50 cm^3 flask preferred to fit apparatus shown in the diagram below
■ measuring cylinder, 25 cm^3	to fit apparatus in diagram below
■ rubber tubing	to fit condenser and receiver in diagram below
■ separating funnel, 50 cm^3, with stopper	
■ thermometer, 0–100 °C	
■ anti-bumping chips	
■ Bunsen burner, tripod and gauze	
■ 3 retort stands, bosses and clamps	
■ safety spectacles and protective gloves	
■ spatula	
■ 2-methylpropan-2-ol, $(CH_3)_3COH$, (tert-butyl alcohol)	~9 cm^3. May need warming (melting point 25 °C) – since students need it in the liquid state
■ sodium sulphate, anhydrous, Na_2SO_4	~ 5 g
■ hydrochloric acid, concentrated, HCl	~ 20 cm^3
■ sodium hydrogencarbonate solution, saturated, $NaHCO_3$	~ 10 cm^3. Shake about 9 g of solid in 100 cm^3 of water. Decant from excess solid after standing
■ balance, sensitivity ± 0.1 g	

HAZARD WARNING 2-Chloro-2-methylpropane and 2-methylpropan-2-ol are flammable. Concentrated hydrochloric acid is corrosive and gives off a harmful vapour. Therefore you **must**:
- **keep stoppers on bottles as much as possible;**
- **keep flammable liquids away from flames;**
- **wear gloves and safety spectacles.**

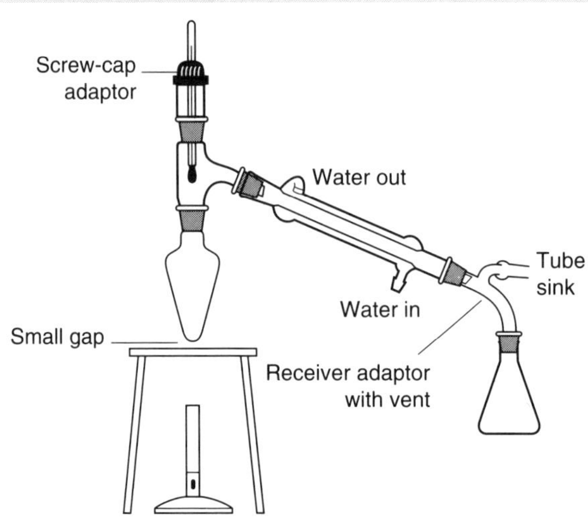

EQUILIBRIUM I: PRINCIPLES

EXPERIMENT 6.1 The effect of concentration changes on equilibria

Requirements per student (or pair)	Notes
■ glass stirring rod	
■ 4 test-tubes	of equal size
■ 2 teat pipettes	
■ safety spectacles	
■ spatula	
■ test-tube rack	
■ ammonium chloride, NH_4Cl	about 2 g
■ wash-bottle of distilled water	
■ iron(III) chloride solution, 0.5 M $FeCl_3$	a few drops. Dissolve 13.5 g of hydrated solid in water, acidify with 12 cm^3 of concentrated hydrochloric acid and make up to 100 cm^3
■ potassium thiocyanate solution, 0.5 M KCNS	a few drops. Dissolve 5 g of solid in water and make up to 100 cm^3

EXPERIMENT 6.2 Determining an equilibrium constant

Requirements per student (or pair)	Notes
Part A	
■ 2 measuring cylinders, 10 cm^3	one must be dry
■ pipette, 5 cm^3	
■ 5 specimen tubes, 75 × 25 mm, with well-fitting caps	
■ access to a balance	sensitivity 0.01 g or better
■ labels for tubes and stoppers	
■ safety pipette filler	
■ distilled water	about 20 cm^3
■ ethyl ethanoate, $CH_3CO_2C_2H_5$	about 20 cm^3
■ hydrochloric acid, 2 M HCl	about 30 cm^3. Add 172 cm^3 of concentrated acid to water and make up to 1000 cm^3
■ safety spectacles	
Part B	
■ burette, 50 cm^3	
■ 5 conical flasks, 250 cm^3	
■ small funnel	for burette filling
■ burette stand	
■ safety spectacles	
■ white tile	
■ wash-bottle of distilled water	
■ sodium hydroxide solution, 1 M NaOH (standardised)	about 150 cm^3. See below
■ phenolphthalein indicator solution	in a dropping bottle. Dissolve 0.1 g of solid in 60 cm^3 of ethanol and make up to 100 cm^3 with distilled water

HAZARD WARNING Concentrated hydrochloric acid and sodium hydroxide are corrosive. Therefore you **must:**

■ **wear gloves and eye protection and follow instructions for dilution carefully.**

1 M NaOH, standardised. Weigh out 40 g of fresh pellets in a closed vessel, dissolve in freshly boiled, cooled, distilled water and make up to 1000 cm^3. Standardise using either standard HCl and methyl orange indicator or potassium hydrogenphthalate and phenolphthalein indicator. Since potassium hydrogenphthalate has limited solubility a 0.1 M solution should be prepared (20.4 g dm^{-3}) and a portion of the 1 M NaOH solution diluted accurately × 10 for the standardisation.

Write the concentration to three significant figures on the label.

Alternatively, ampoules, which can be diluted carefully to make precisely 1 M NaOH, are available from most chemical suppliers and are very convenient, if rather expensive.

EXPERIMENT 6.3 Determining a solubility product

Requirements per student (or pair)	Notes
■ 4 reagent bottles, 250 cm^3	with stoppers
■ burette, 50 cm^3	
■ 4 conical flasks, 250 cm^3	
■ 4 filter funnels	must be dry
■ small funnel	for burette filling
■ measuring cylinder, 100 cm^3	
■ pipette, 25 cm^3	
■ filter papers	
■ labels for bottles	
■ burette stand	
■ safety spectacles	
■ safety pipette filler	
■ spatula	
■ thermometer, 0–100 °C (±1 °C)	
■ white tile	
■ calcium hydroxide, $Ca(OH)_2$	at least 10 g
■ distilled water	at least 500 cm^3
■ hydrochloric acid, 0.1 M HCl (standardised)	about 60 cm^3. See below
■ phenolphthalein indicator solution	in a dropping bottle

HAZARD WARNING Concentrated hydrochloric acid is corrosive. Therefore you **must**:

■ **wear gloves and eye protection and follow instructions for dilution carefully.**

0.1 M HCl, standardised. Add 8.6 cm^3 (or 10.3 g) of concentrated acid to distilled water and make up to 1000 cm^3 in a volumetric flask. Standardise using 0.0500 M sodium carbonate solution (5.30 g of anhydrous Na_2CO_3 per dm^3) and methyl orange indicator. Write the concentration to three significant figures on the label. Alternatively, ampoules, which can be diluted carefully to make precisely 0.1 MHCl, are available from most chemical suppliers and are very convenient, if rather expensive.

EXPERIMENT 6.4 Illustrating the common ion effect

Requirements per student (or pair)	Notes
■ teat pipette	
■ 2 test-tubes with corks	
■ safety spectacles	
■ spatula or forceps	
■ test-tube rack	
■ sodium hydroxide, NaOH	a few pellets in a stoppered bottle
■ hydrochloric acid, concentrated, HCl	in a dropping bottle
■ sodium chloride solution, saturated	about 20 cm^3. Stir about 40 g of salt with 100 cm^3 of warm water and allow to cool. Leave excess solid in bottom of stock bottle

HAZARD WARNING

Concentrated hydrochloric acid and sodium hydroxide are corrosive. Therefore you **must**:

■ **wear gloves and eye protection and follow instructions for dilution carefully.**

EXPERIMENT 6.5 Distribution equilibrium

Requirements per student (or pair)	Notes
■ 2 beakers, 100 cm^3	
■ burette, 50 cm^3, and stand	
■ 2 conical flasks, 150 cm^3	
■ measuring cylinder, 50 cm^3	
■ pipette, 10 cm^3	
■ separating funnel, 150 cm^3	
■ safety spectacles	
■ safety pipette filler	
■ white tile	
■ wash-bottle of distilled water	
■ ammonia solution, ~ 1 M NH_3	about 50 cm^3. See below
■ hydrochloric acid, 0.5 M HCl (standardised)	about 200 cm^3. See below
■ hydrochloric acid, 0.01 M HCl (standardised)	about 200 cm^3. See below
■ Volasil 244*	about 50 cm^3
■ methyl red indicator solution (or methyl orange)	in a dropping bottle. See below
■ Bottle labelled 'organic residues' in fume cupboard	

HAZARD WARNING 0.880 Ammonia and concentrated hydrochloric acid are corrosive and give off harmful vapours. Therefore you **must**:

■ **wear gloves and eye protection, and use a fume cupboard when diluting.**

Volasil 244 is flammable. Therefore you **must**:

■ **keep away from flames and replace the stopper.**

Approx. 1 M ammonia solution. Add 56 cm^3 of 0.880 ammonia to water, and make up to 1 dm^3. Avoid breathing fumes, and any risk of ammonia getting into eyes.
0.5 M HCl (standardised). Transfer 43 cm^3 of concentrated acid to water and make up to 1 dm^3. Standardise with sodium carbonate solution, 0.250 M Na_2CO_3 (26.5 g of anhydrous sodium carbonate per dm^3) and methyl orange indicator.
0.01 M HCl (standardised). Transfer, by pipette, 20.0 cm^3 of 0.5 M HCl to a volumetric flask and make up to 1000 cm^3. Alternatively, make the dilution approximately and standardise using 0.0100 M Na_2CO_3 (1.059 g of AR anhydrous salt per dm^3) and methyl orange indicator.
Methyl red indicator solution. Dissolve 0.02 g of dye in 20 cm^3 of ethanol and make up to 100 cm^3 with distilled water.
Methyl orange indicator solution. Dissolve 0.04 g of dye in 20 cm^3 of ethanol and make up to 100 cm^3 with distilled water.

*Volasil 244 is a cyclic organosilicone, octamethylcyclotetrasiloxane. Its molecule has alternating silicon and oxygen atoms in a ring structure with methyl groups attached to the silicon atoms. We have included it here, and subsequently, because it is more environmentally friendly than most organic solvents and is safer to use, although it is flammable. Volasil 244 may be obtained from BDH Laboratory Supplies, Merck Ltd, Hunter Boulevard, Lutterworth, LE17 4XN. Tel: 0800 223344. (1997 prices: £12.80 for 500 cm^3, £27.10 for 2.5 litres. Catalogue number 63300.)

ILPAC 7

EQUILIBRIUM II: ACIDS AND BASES

TECHNICIAN'S SHEET

EXPERIMENT 7.1 The pH of a weak acid at various concentrations

Requirements per student (or pair)	Notes
■ beaker, 50 cm^3	
■ wash-bottle of distilled water	
■ pH meter and electrode	
■ ethanoic acid, 0.10 M CH_3CO_2H	~25 cm^3. Dissolve 5.8 cm^3 (or 6.0 g) of pure ethanoic acid (glacial acetic acid) in distilled water and make up to 1 dm^3
■ ethanoic acid, 0.010 M CH_3CO_2H	~25 cm^3. Measure 1 volume of 0.10 M solution and make up to 10 volumes
■ ethanoic acid, 0.0010 M CH_3CO_2H	~25 cm^3. Measure 1 volume of 0.010 M solution and make up to 10 volumes
■ ethanoic acid, 0.00010 M CH_3CO_2H	~25 cm^3. Measure 1 volume of 0.0010 M solution and make up to 10 volumes
■ buffer solution, pH 4.0	~25 cm^3. Either follow the instructions supplied with commercial buffer tablets, or use one of the recipes below

HAZARD WARNING

Concentrated ethanoic acid is corrosive. Therefore you **must:**

■ **wear safety spectacles.**

Buffer solution, pH 4.0

Either dissolve 12.91 g of 2-hydroxypropane-1,2,3-tricarboxylic acid-1-water (citric acid), $C_6H_8O_7{\cdot}H_2O$ and 10.95 g of anhydrous disodium hydrogen phosphate, Na_2HPO_4, in distilled water and make up to 1 dm^3. If you have only the hydrated salt, $Na_2HPO_4{\cdot}12H_2O$, use 27.6 g.

or 0.05 M potassium hydrogenphthalate. Dissolve 10 g of potassium hydrogenphthalate in distilled water and make up to 1 dm^3.

EXPERIMENT 7.2 The pH of different acids at the same concentration

Requirements per student (or pair)	Notes
■ 2 beakers, 50 cm^3	
■ wash-bottle of distilled water	
■ pH meter and electrode	
■ benzoic acid solution, 0.010 M $C_6H_5CO_2H$	~25 cm^3. Dissolve 1.22 g of solid in about 800 cm^3 of hot water. Cool and make up to 1 dm^3
■ boric acid solution, 0.010 M H_3BO_3	~25 cm^3. Dissolve 0.62 g of solid in water and make up to 1 dm^3
■ dihydrogenphosphate(V) ion solution, 0.010 M $H_2PO_4^-$	~25 cm^3. Dissolve 1.56 g of solid sodium dihydrogenphosphate-2-water, $NaH_2PO_4{\cdot}2H_2O$ (or 1.36 g of KH_2PO_4) in water and make up to 1 dm^3
■ ethanoic acid solution, 0.010 M CH_3CO_2H	~25 cm^3. Dissolve 0.58 cm^3 (or 0.60 g) of pure ethanoic acid (glacial acetic acid) in water and make up to 1 dm^3
■ buffer solution, pH 4.0	~25 cm^3. Either follow the instructions supplied with commercial buffer tablets, or use one of the recipes below

HAZARD WARNING

Concentrated ethanoic acid is corrosive. Therefore you **must:**

- **wear safety spectacles.**

Buffer solution, pH 4.0

Either dissolve 12.91 g of 2-hydroxypropane-1,2,3-tricarboxylic acid-1-water (citric acid), $C_6H_8O_7{\cdot}H_2O$ and 10.95 g of anhydrous disodium hydrogen phosphate, Na_2HPO_4, in distilled water and make up to 1 dm^3. If you have only the hydrated salt, $Na_2HPO_4{\cdot}12H_2O$, use 27.6 g.

or 0.05 M potassium hydrogenphthalate. Dissolve 10 g of potassium hydrogenphthalate in distilled water and make up to 1 dm^3.

EXPERIMENT 7.3 The action of a buffer solution

Requirements per student (or pair)	Notes
■ beaker, 50 cm^3	
■ 2 beakers, 100 cm^3	
■ 2 burettes	these may be shared
■ 2 funnels, small	for filling burettes
■ measuring cylinder, 25 cm^3	
■ 2 burette stands	
■ pH meter and electrode	if no pH meter is available use: ■ thin glass stirring rod ■ full range pH paper, pH 1–14 ■ narrow range pH papers to cover pH 2–12
■ safety spectacles	
■ wash-bottle of distilled water	
■ pure water	~25 cm^3. This must be **freshly** deionised, or **freshly** boiled distilled water, and stoppered securely just before the lesson to prevent absorption of CO_2
■ hydrochloric acid, 0.1 M HCl	~10 cm^3. Add 8.6 cm^3 of concentrated HCl to water and make up to 1 dm^3
■ sodium hydroxide solution, 0.1 M NaOH	~10 cm^3. Dissolve 4.0 g of NaOH pellets in water and make up to 1 dm^3
■ buffer solution, pH 7	~25 cm^3. *Either* follow the instructions supplied with commercial tablets *or* use the recipe below

HAZARD WARNING

Concentrated hydrochloric acid is a corrosive liquid, and its vapour is harmful to eyes, lungs and skin.

Sodium hydroxide solution is also corrosive. Therefore you **must**:
- **wear safety spectacles.**

Buffer solution, pH 7 Dissolve **one** of the dihydrogenphosphates from list A and **one** of the hydrogenphosphates from list B in distilled water and make up to 1 dm^3.

A		B	
$NaH_2PO_4{\cdot}2H_2O$	6.08 g	$Na_2HPO_4{\cdot}12H_2O$	21.85 g
KH_2PO_4	5.31 g	Na_2HPO_4	8.66 g
		K_2HPO_4	10.62 g

EXPERIMENT 7.4 Determining the pH range of some acid–base indicators

Requirements per student (or pair)	Notes
■ 3 beakers, 100 cm^3	
■ 4 beakers, 250 cm^3	
■ 3 burettes	2 may be shared
■ 3 funnels, small	to fill burettes
■ pipette, 25 cm^3	
■ 3 burette stands	
■ pH meter and electrode	if no meter is available, students can calculate the pH
■ pipette filler	
■ safety spectacles	
■ wash-bottle of distilled water	
■ hydrochloric acid, 0.1 M HCl	~125 cm^3. Add 8.6 cm^3 of concentrated HCl to water and make up to 1 dm^3
■ sodium hydroxide solution, 0.10 M NaOH	~125 cm^3. Dissolve 4.00 g of NaOH pellets in water and make up to 1 dm^3. If this is done carefully, the solution need not be standardised
■ buffer solution, pH 3.1 (or 3.25)	~125 cm^3. Follow the instructions supplied with BDH universal buffer solution, which has an unpleasant smell, or use the recipe below
■ litmus indicator	dissolve 1.0 g of solid in boiling water, filter and make up to 100 cm^3
■ bromophenol blue indicator	dissolve in ethanol: 0.04 g + 20 cm^3 C_2H_5OH
■ methyl orange indicator	0.04 g + 20 cm^3 C_2H_5OH
■ methyl red indicator	0.02 g + 20 cm^3 C_2H_5OH
■ phenolphthalein indicator	0.10 g + 60 cm^3 C_2H_5OH

(Indicator solutions: Make up to 100 cm^3 with H_2O)

HAZARD WARNING

Concentrated hydrochloric acid is a corrosive liquid, and its vapour is harmful to eyes, lungs and skin.

Sodium hydroxide solution is also corrosive. Therefore you **must**:

- **wear safety spectacles.**

Universal buffer solution, pH 3.25 Dissolve the following materials in distilled water and make up to 1 dm^3:

Boric acid, H_3BO_3	2.47 g
Potassium dihydrogenphosphate, KH_2PO_4	5.44 g
(*or* sodium dihydrogenphosphate-2-water, $NaH_2PO_4{\cdot}2H_2O$)	(6.24 g)
Ethanoic acid, CH_3CO_2H, (glacial acetic acid)	2.40 g
(*or* ethanoic acid solution, 1.00 M CH_3CO_2H)	(40.0 cm^3)

EXPERIMENT 7.5 Determining the ionisation constant for an indicator

Requirements per student (or pair)	Notes
■ measuring cylinder, 10 cm^3	
■ stirring rod	
■ 2 teat pipettes	
■ 20 test-tubes	all the same size
■ test-tube rack or racks	to hold 2 rows of 9 tubes
■ safety spectacles	
■ wash-bottle of distilled water	
■ bromophenol blue solution	~10 cm^3. Dissolve 0.04 g of solid in 20 cm^3 of ethanol and make up to 100 cm^3 with distilled water
■ buffer solution, pH 3.7	~10 cm^3. See below
■ hydrochloric acid, concentrated, HCl	a few drops
■ sodium hydroxide solution, 4 M NaOH	a few drops. Carefully dissolve 16 g of NaOH pellets in water and make up to 100 cm^3

HAZARD WARNING

Concentrated hydrochloric acid is a corrosive liquid, and its vapour is harmful to eyes, lungs and skin.

Sodium hydroxide solution is also corrosive. Therefore you **must**:

■ **wear safety spectacles.**

Buffer solution, pH 3.7 Dissolve 24.31 g of disodium hydrogenphosphate(V)-12-water, $Na_2HPO_4{\cdot}12H_2O$, (*or* 9.62 g of anhydrous Na_2HPO_4, *or* 11.79 g of K_2HPO_4) and 6.94 g of 2-hydroxypropane-1,2,3-tricarboxylic acid-1-water (citric acid), $C_3H_4OH(CO_2H)_3{\cdot}H_2O$ in water and make up to 1 dm^3.

EQUILIBRIUM II: ACIDS AND BASES

EXPERIMENT 7.6 Obtaining the pH curve for an acid–alkali titration

Requirements per student (or pair)	Notes
■ conical flask, 250 cm^3	
■ 2 beakers, 250 cm^3	
■ burette, 50 cm^3	
■ funnel, small	
■ beaker, tall form, 100 or 150 cm^3	
■ pipette, 25 cm^3	
■ burette stand	
■ 2 labels for beakers	
■ pH meter and electrode	if no pH meter is available use: ■ thin glass stirring rod ■ pH paper, pH 1–14 ■ narrow range pH papers to cover pH 1–14
■ pipette filler	
■ safety spectacles	
■ stirrer, magnetic or electric	
■ wash-bottle of distilled water	
■ phenolphthalein indicator solution	see below
■ ethanoic (acetic) acid, ~0.1 M CH_3CO_2H	~100 cm^3
■ phenol red indicator solution	labelled 'A'. See below
■ sodium hydroxide solution, ~0.1 M NaOH	~100 cm^3

HAZARD WARNING

Sodium hydroxide solution is very corrosive. Even when dilute it can damage your eyes. Therefore you **must**:

■ **wear safety spectacles throughout the experiment.**

Phenolphthalein indicator solution Dissolve 1 g of the reagent in 100 cm^3 of ethanol (IMS) and add 100 cm^3 of distilled water with constant stirring. Filter if a precipitate forms. Alternatively, the indicator solution can be bought already prepared for use.

Phenol red indicator solution If the solid indicator is in the acid form, put 0.1 g of phenol red in a small beaker and add 2.85 cm^3 of sodium hydroxide solution, 0.1 M NaOH. Use a glass rod with a flattened end to grind and stir (triturate) to dissolve the solid. Dilute to 100 cm^3 with distilled water. Filter if necessary. If the water-soluble form is available, dissolve 0.1 g in 100 cm^3 of distilled water. Alternatively, the indicator solution can be bought already prepared for use.

EQUILIBRIUM II: ACIDS AND BASES

EXPERIMENT 7.7 Preparation of buffers

Since this experiment is designed as a planning exercise, students will probably be asked to submit their own list of requirements, but the following list should be adequate. The experiment is similar to Experiment 7.3.

Requirements per student (or pair)	Notes
■ 2 beakers, 100 cm^3	
■ beaker, 50 cm^3	
■ 2 burettes, 50 cm^3	
■ 2 funnels, small	to fill burettes
■ measuring cylinder, 25 cm^3	
■ stirring rod, thin glass	
■ 2 burette stands	
■ pH meter	
■ pH papers, full range	required only if pH meter is not available
■ pH papers, narrow range	
■ safety spectacles	
■ wash-bottle of distilled water	
■ ammonia solution, 1.0 M NH_3	~30 cm^3. Carefully add 56 cm^3 of 0.880 ammonia to water and make up to 1 dm^3. This solution should preferably be standardised. Alternatively, make up from commercial ampoule by accurate dilution
■ ammonium chloride solution, 1.0 M NH_4Cl	~80 cm^3. Dissolve 53.5 g of solid in water and make up to 1 dm^3
■ ethanoic (acetic) acid, 1.0 M CH_3CO_2H	~30 cm^3. Add 58 cm^3 of glacial acid to water and make up to 1 dm^3. This solution should preferably be standardised. Alternatively, make up from commercial ampoules by accurate dilution
■ sodium ethanoate (acetate) solution, 1.0 M CH_3CO_2Na	~80 cm^3. Dissolve 82.03 g of the anhydrous salt, or 136.1 g of the trihydrate, in water and make up to 1 dm^3

HAZARD WARNING

Ammonia solution is harmful and the vapour is irritating to the eyes and lungs. Therefore you **must**:

- **wear safety spectacles;**
- **prepare the ammonia solution at the fume cupboard.**

EQUILIBRIUM II: ACIDS AND BASES

TECHNICIAN'S SHEET

EXPERIMENT 7.8 Determining the dissociation constant of a weak acid

Requirements per student (or pair)	Notes
■ measuring cylinder, 10 cm^3	
■ stirring rod	
■ 2 teat pipettes	
■ 20 test-tubes	all the same size
■ test-tube rack or racks	to hold 2 rows of 9 tubes
■ safety spectacles	
■ protective gloves	
■ wash-bottle of distilled water	
■ bromophenol blue solution	~10 cm^3. Dissolve 0.04 g of solid in 20 cm^3 of ethanol and make up to 100 cm^3 with distilled water
■ benzoic acid solution, 0.02 M $C_6H_5CO_2H$	~5 cm^3. Dissolve 2.44 g of solid in 800 cm^3 of hot water, cool and make up to 1 dm^3
■ sodium benzoate solution, 0.02 M $C_6H_5CO_2Na$	~5 cm^3. Dissolve 2.88 g of anhydrous solid in water and make up to 1 dm^3
■ hydrochloric acid, concentrated, HCl	a few drops
■ sodium hydroxide solution, 4 M NaOH	a few drops. Carefully dissolve 16 g of NaOH pellets in water and make up to 100 cm^3

HAZARD WARNING

Concentrated hydrochloric acid is a corrosive liquid, and its vapour is harmful to eyes, lungs and skin.

Sodium hydroxide solution is also corrosive. Therefore you **must**:

■ **wear safety spectacles.**

ILPAC 7

EQUILIBRIUM III: REDOX REACTIONS

TECHNICIAN'S SHEET

EXPERIMENT 7.9 Some simple redox reactions

Requirements per student (or pair)	Notes
■ 6 test-tubes	
■ emery paper	for cleaning metal strips
■ safety spectacles	
■ test-tube rack	
■ copper, foil, Cu	two strips, 5 cm × 0.5 cm
■ silver, wire, Ag	two lengths, 5 cm
■ zinc, foil, Zn	two strips, 5 cm × 0.5 cm
■ copper(II) sulphate solution, 0.50 M $CuSO_4$	~10 cm^3. Dissolve 12.5 g of $CuSO_4{\cdot}5H_2O$ in water and make up to 100 cm^3
■ silver nitrate solution, 0.10 M $AgNO_3$	~10 cm^3. Dissolve 1.7 g of solid in water and make up to 100 cm^3. Store and dispense in an amber bottle
■ zinc sulphate solution, 0.50 M $ZnSO_4$	~10 cm^3. Dissolve 14.5 g of $ZnSO_4{\cdot}7H_2O$ in water and make up to 100 cm^3
■ silver residues bottle	see below

HAZARD WARNING

Copper sulphate is harmful.
Silver nitrate is corrosive and **toxic**. Therefore you **must**:
■ **wear safety spectacles.**

Silver residues In view of the high cost of silver nitrate, we recommend that all residues should be collected in a large amber bottle. Concentrated hydrochloric acid should be added from time to time to ensure complete precipitation as silver chloride. When the precipitate has settled, the supernatant liquid may be poured off to make room for fresh residues. When sufficient silver chloride has been collected in this way, it may be sold to a chemical supplier, or recovered as silver nitrate performed in the laboratory. (See a suitable laboratory manual for details.)

EQUILIBRIUM III: REDOX REACTIONS

EXPERIMENT 7.10 Measuring the potential difference generated by electrochemical cells

Requirements per student (or pair)	Notes
■ 4 beakers, 50 cm^3	
■ emery paper	
■ 6 filter paper strips	~10 cm × 1 cm
■ safety spectacles	
■ voltmeter, high resistance, 2 V	most pH meters double as valve voltmeters. Alternatively, use a modern AVO meter with input impedance $10^4\ \Omega$
■ 3 connecting leads	one end to be connected to voltmeter, the other to a metal strip via a crocodile clip
■ copper, foil, Cu	one strip, 5 cm × 0.5 cm
■ silver, wire, Ag	one piece, 5 cm
■ zinc, foil, Zn	one strip, 5 cm × 0.5 cm
■ copper(II) sulphate solution, 1.0 M $CuSO_4$	~20 cm^3. Dissolve 250 g of $CuSO_4{\cdot}5H_2O$ in water and make up to 1000 cm^3
■ potassium nitrate solution, saturated, 3 M KNO_3	~20 cm^3. Dissolve 300 g of solid in water and make up to 1000 cm^3
■ silver nitrate solution, 0.10 M $AgNO_3$	~20 cm^3. See Experiment 7.9
■ zinc sulphate solution, 1.0 M $ZnSO_4$	~20 cm^3. Dissolve 288 g of $ZnSO_4{\cdot}7H_2O$ in water and make up to 1000 cm^3
■ silver residues bottle	see Experiment 7.9

HAZARD WARNING

Copper sulphate is harmful.
Silver nitrate is corrosive and **toxic**. Therefore you **must**:
■ **wear safety spectacles.**

EQUILIBRIUM III: REDOX REACTIONS

EXPERIMENT 7.11 Testing predictions about redox reactions

Requirements per student (or pair)	Notes
■ 10 test-tubes, with corks or bungs to fit	
■ safety spectacles	
■ spatula	
■ test-tube rack	
■ copper, powder, Cu	~1 g
■ Volasil 244*	~2 cm^3
■ zinc, powder, Zn	~1 g
■ bromine water, $Br_2(aq)$	~6 cm^3. Carefully break a 1 cm^3 phial under 100 cm^3 of water and shake to dissolve
■ chromium(III) chloride solution, 0.10 M $CrCl_3$	~3 cm^3. Dissolve 2.7 g of $CrCl_3{\cdot}6H_2O$ in water and make up to 100 cm^3
■ iron(III) chloride solution, 0.10 M $FeCl_3$	~3 cm^3. Dissolve 1.6 g of $FeCl_3$ in water and make up to 100 cm^3
■ potassium bromide solution, 0.10 M KBr	~3 cm^3. Dissolve 1.2 g of KBr in water and make up to 100 cm^3
■ potassium chloride solution, 0.10 M KCl	~3 cm^3. Dissolve 0.8 g of KCl in water and make up to 100 cm^3
■ potassium hexacyanoferrate(III) solution (ferricyanide), 0.10 M $K_3Fe(CN)_6$	~1 cm^3. Dissolve 3.3 g of $K_3Fe(CN)_6$ in water and make up to 100 cm^3
■ potassium iodide solution, 0.10 M KI	~3 cm^3. Dissolve 1.7 g of KI in water and make up to 100 cm^3
■ potassium manganate(VII) solution (permanganate), 0.020 M $KMnO_4$	~6 cm^3. Dissolve 0.32 g of $KMnO_4$ in water and make up to 100 cm^3
■ potassium peroxodisulphate solution (persulphate), 0.10 M $K_2S_2O_8$	~3 cm^3. Dissolve 2.7 g of $K_2S_2O_8$ in water and make up to 100 cm^3
■ sulphuric acid, dilute, 1 M H_2SO_4	~2 cm^3. Carefully add 55 cm^3 of concentrated acid, slowly and with constant stirring, to 800 cm^3 of water and make up to 1000 cm^3

HAZARD WARNING

Bromine vapour must not be inhaled. If a fume cupboard is not available then:
- **the laboratory must be well ventilated and reagent bottles and test-tubes stoppered as much as possible.**

Volasil 244 is flammable. Therefore you **must**:
- **keep Volasil 244 away from flames.**

Potassium hexacyanoferrate(III) is **toxic**. Bromine vapour and all the chemical solutions used in this experiment are irritants. Therefore you **must**:
- **wear safety spectacles.**

*Volasil 244 is a cyclic organosilicone, octamethylcyclotetrasiloxane. Its molecule has alternating silicon and oxygen atoms in a ring structure with methyl groups attached to the silicon atoms. We have included it here, and subsequently, because it is more environmentally friendly than most organic solvents and is safer to use, although it is flammable. Volasil 244 may be obtained from BDH Laboratory Supplies, Merck Ltd, Hunter Boulevard, Lutterworth, LE17 4XN. Tel: 0800 223344. (1997 prices: £12.80 for 500 cm^3, £27.10 for 2.5 litres. Catalogue number 63300.)

EQUILIBRIUM III: REDOX REACTIONS

TECHNICIAN'S SHEET

EXPERIMENT 7.12 Variation of cell e.m.f. with concentration

Requirements per student (or pair)	Notes
■ 5 beakers, 50 cm^3	
■ 3 filter paper strips	~10 cm × 1 cm
■ safety spectacles	
■ voltmeter, high resistance, 2 V	see Experiment 7.10
■ 2 connecting leads	one end to voltmeter, the other with a crocodile clip
■ copper, foil, Cu	one strip, 5 cm × 0.5 cm
■ zinc, foil, Zn	one strip, 5 cm × 0.5 cm
■ emery paper	
■ potassium nitrate solution, 3 M KNO_3	~20 cm^3. See Experiment 7.10
■ copper(II) sulphate solution, 1.0 M $CuSO_4$	~20 cm^3. See Experiment 7.10
■ copper(II) sulphate solution, 0.10 M $CuSO_4$	~20 cm^3. Accurately dilute 1 volume of 1.0 M $CuSO_4$ with 9 volumes of water
■ zinc sulphate solution, 1.0 M $ZnSO_4$	~20 cm^3. See Experiment 7.10
■ zinc sulphate solution, 0.10 M $ZnSO_4$	~20 cm^3. Accurately dilute 1 volume of 1.0 M $ZnSO_4$ with 9 volumes of water

HAZARD WARNING

Copper sulphate is harmful.
Therefore you **must**:
■ **wear safety spectacles**.

EXPERIMENT 7.13 Rusting of iron and steel

Requirements per student (or pair)	Notes
■ safety spectacles	
■ Petri or crystallising dish	must be flat-bottomed ≥7 cm diameter
■ beaker, 250 cm^3	
■ dropping pipette	
■ emery paper	
■ piece of steel sheet or plate	preferably at least 2 cm × 2 cm
■ 5 iron nails	oval wire nails, 40 mm or 50 mm, obtainable from a DIY store
■ zinc foil	1 strip, 1.5 cm × 0.5 cm
■ copper foil	1 strip, 1.5 cm × 0.5 cm
■ Vaseline or stopcock grease	
■ piece of cotton wool or cloth	
■ methanol	
■ phosphoric acid, 1 M H_3PO_4	add 70 cm^3 of syrupy phosphoric acid slowly with stirring to 800 cm^3 of distilled water and make up to 1000 cm^3
■ ferroxyl indicator	prepare fresh. Dissolve 3 g of potassium hexacyanoferrate(III) (potassium ferricyanide) $K_3Fe(CN)_6$ and 20 g of sodium chloride in 500 cm^3 of distilled water. Stir in 1 cm^3 of phenolphthalein indicator

HAZARD WARNING

Ferroxyl indicator contains potassium hexacyanoferrate(III) which is poisonous and boiling dilute phosphoric acid solution is hazardous. Therefore you **must**:

- **wear safety spectacles.**

EXPERIMENT 7.14 Determination of the Faraday constant

Requirements per student (or pair)	Notes
■ d.c. power source (low voltage)	
■ ammeter, 0–1 A	the power source may contain this
■ rheostat	the power source may contain this
■ connecting wire	
■ 2 connecting leads	one end to power source, the other with a crocodile clip
■ 2 pieces of copper foil or plate	6 cm × 3 cm
■ emery paper	
■ beaker, 150 or 250 cm^3	
■ glass rod	
■ stopclock	
■ balance	capable of weighing to at least 0.01 g and preferably to 0.001 g
■ copper(II) sulphate, 0.5 M $CuSO_4$	100 cm^3. See Experiment 7.9
■ propanone (acetone)	
■ wash-bottle of distilled water	

HAZARD WARNING

Copper sulphate is harmful. Propanone is flammable. Therefore you **must**:

- **wear safety spectacles;**
- **keep propanone away from flames.**

EXPERIMENT 8.1 Chemical properties of ethanol – Parts A and B

Requirements per student (or pair)	Notes
Part A	
■ test-tube	
■ safety spectacles	
■ wash-bottle of distilled water	
■ ethanol	~1 cm^3. IMS is suitable
■ universal indicator solution	in dropping bottle
Part B	
■ beaker, 250 cm^3	
■ boiling tube	
■ funnel, small, wide stem	for adding solid to flask
■ ground-glass-joint apparatus for distillation	as shown in the diagram below. Rubber tubing also required for condenser and receiver. If a receiver adaptor with a vent is not available, then an ordinary receiver adaptor can be used with the apparatus at a fume cupboard or in a well-ventilated room
■ measuring cylinder, 10 cm^3	
■ 3 teat pipettes	
■ test-tube and holder	
■ anti-bumping granules	
■ balance, sensitivity ±0.1 g	
■ Bunsen burner, tripod, gauze and bench mat	
■ 2 retort stands, bosses and clamps	
■ safety spectacles	
■ protective gloves	
■ spatula	
■ ethanol, C_2H_5OH	~5 cm^3. IMS is suitable
■ sodium dichromate(VI), $Na_2Cr_2O_7$	~3 g
■ ammonia solution, 2 M NH_3	~2 cm^3. See note below
■ Fehling's solutions, 1 and 2	~1 cm^3 of each. See note below
■ silver nitrate solution, 0.05 M $AgNO_3$	~5 cm^3. See note below
■ sodium hydroxide solution, 2 M NaOH	~1 cm^3. See note below
■ sulphuric acid, dilute, 1 M H_2SO_4	~10 cm^3. See note below

Note: Instructions for the preparation of the solutions are summarised on pages 127 to 128. For disposal of residues see page 129.

(Please see hazard warnings on page 187, which refer to the whole of Experiment 8.1.)

EXPERIMENT 8.1 Chemical properties of ethanol – Parts A and B (continued)

HAZARD WARNING

Ethanol is very flammable. Therefore you **must**:

- **keep the stopper on the bottle as much as possible;**
- **keep the bottle away from flames;**
- **wear safety spectacles.**

Bromine and glacial ethanoic acid (acetic acid) have dangerous fumes and burn the skin. Therefore you **must**:

- **keep the stoppers on the bottles as much as possible;**
- **wear gloves;**
- **wear safety spectacles.**

Concentrated sulphuric acid is very corrosive and reacts violently with water. Therefore you **must**:

- **wear gloves;**
- **wear safety spectacles;**
- **mop up minor spillages with plenty of water;**
- **dispose of unwanted residues by cooling and pouring slowly into an excess of water.**

Tollens reagent for the silver mirror test becomes explosive if dry. Therefore you **must**:

- **wash away the solutions from experiments A and B immediately you have finished.**

Sodium is extremely dangerous. Therefore you **must**:

- **use as little as possible (1 mm cube);**
- **use forceps to handle it;**
- **keep it under oil;**
- **wear safety spectacles.**

Sodium dichromate(VI) is a powerful oxidant and can damage the skin. Therefore you **must**:

- **wear safety spectacles and gloves.**

Sodium hydroxide solution is very corrosive. Even when dilute it can damage your eyes. Therefore you **must:**

- **wear safety spectacles.**

Fehling's solution 2 is corrosive because it contains sodium hydroxide. It is likely to spurt out of a clean test-tube during heating. Therefore you **must:**

- **wear safety spectacles.**

EXPERIMENT 8.1 Chemical properties of ethanol – Parts C and D

Requirements per student (or pair)	Notes
Part C	
■ funnel, small, wide stem	
■ ground-glass-joint apparatus for distillation	as shown below. Rubber tubing also required for condenser and receiver. If a receiver adaptor with a vent is not available, then an ordinary receiver adaptor can be used with the apparatus at a fume cupboard or in a well-ventilated room
■ measuring cylinder, 10 cm^3	
■ teat pipette	
■ test-tube	
■ balance, sensitivity ±0.1 g	
■ Bunsen burner, tripod, gauze and bench mat	
■ 2 retort stands, bosses and clamps	
■ safety spectacles and protective gloves	
■ spatula	
■ universal indicator papers	
■ anti-bumping granules	
■ ethanol, C_2H_5OH	~1 cm^3. IMS is suitable
■ sodium carbonate, Na_2CO_3	~1 g (anhydrous)
■ sodium dichromate(VI), $Na_2Cr_2O_7$	~5 g
■ sulphuric acid, concentrated, H_2SO_4	~2 cm^3. Small bottle with dropper placed on a drip tray
■ sulphuric acid, dilute, 1 M H_2SO_4	~10 cm^3. See note below
Part D	
■ test-tube and rack	
■ watch glass	
■ filter papers	for removing oil from sodium
■ forceps	for handling sodium
■ safety spectacles	
■ wood splint	
■ ethanol, C_2H_5OH	~1 cm^3. IMS is suitable
■ sodium metal, Na	two 1 mm cubes under the same oil as is used for storage

Note: Instructions for the preparation of the solutions are summarised on pages 127 to 128. For disposal of residues see page 129.

(Please see hazard warnings on page 187, which refer to the whole of Experiment 8.1.)

EXPERIMENT 8.1 Chemical properties of ethanol – Parts E and F

Requirements per student (or pair)	Notes
Part E	
■ beaker, 100 cm^3	
■ teat pipette	
■ test-tube and test-tube holder	
■ Bunsen burner and bench mat	
■ safety spectacles	
■ ethanol, C_2H_5OH	~2 cm^3
■ ethanoic (acetic) acid, glacial, CH_3CO_2H	~1 cm^3
■ sulphuric acid, concentrated, H_2SO_4	~1 cm^3. Small bottle with dropper placed on a drip tray
■ sodium carbonate solution, 1 M Na_2CO_3	~20 cm^3. Dissolve 106 g of anhydrous Na_2CO_3 in water and make up to 1000 cm^3
Part F	
■ conical flask, 100 cm^3, fitted with delivery tubes	see diagram below
■ fume cupboard, and safety spectacles	
■ teat pipette	
■ 3 test-tubes with corks	
■ water-trough or large beaker	see diagram below
■ Bunsen burner and bench mat	
■ ceramic wool	see diagram below
■ 2 retort stands, bosses and clamps	
■ ethanol, C_2H_5OH	~2 cm^3
■ pumice stone, 4–8 mesh	see diagram below
■ bromine water, $Br_2(aq)$	~1 cm^3
■ potassium manganate(VII) (permanganate) solution, 0.01 M $KMnO_4$	~1 cm^3
■ sulphuric acid, dilute, 1 M H_2SO_4	~2 cm^3

Note: For the preparation of the solutions, see pages 127 to 128. For disposal of residues see page 129.

(Please see hazard warnings on page 187, which refer to the whole of Experiment 8.1.)

EXPERIMENT 8.2 Chemical properties of phenol

Requirements per student (or pair)	Notes
■ 2 beakers, 250 cm^3	
■ teat pipette	
■ 6 test-tubes, with corks	
■ thermometer, 0–100 °C	
■ Bunsen burner, tripod, gauze and bench mat	
■ filter papers	for removing oil from sodium
■ forceps	for handling sodium
■ safety spectacles	
■ protective gloves	disposable gloves preferred
■ spatula	
■ wood splint	
■ test-tube holder	
■ test-tube rack	
■ wash-bottle of distilled water	
■ hydrochloric acid, concentrated, HCl	~2 cm^3. Small bottle with dropper placed on a drip tray
■ phenol, C_6H_5OH	~12 g
■ sodium metal, Na	two 1 mm cubes, under the same oil as is used for storage
■ sodium hydrogencarbonate, $NaHCO_3$	~1 g
■ universal indicator solution	in dropping bottle
■ bromine water, $Br_2(aq)$	~1 cm^3. See note below
■ iron(III) chloride solution, 0.5 M $FeCl_3$	~1 cm^3. Dissolve 13 g of solid in water with 2 cm^3 of concentrated hydrochloric acid and make up to 100 cm^3
■ sodium carbonate solution, 1 M Na_2CO_3	~2 cm^3. See Experiment 8.1 – Parts E and F
■ sodium hydroxide solution, 2 M NaOH	~5 cm^3. See note below

HAZARD WARNING

Both phenol and bromine water attack the skin and give off irritating vapours. Therefore you **must**:

- **wear gloves and safety spectacles;**
- **avoid contact with skin;**
- **keep stoppers on bottles as much as possible.**

Sodium is dangerously reactive. Therefore you **must**:

- **use as little as possible (1 mm cube);**
- **handle with forceps;**
- **keep it under oil;**
- **wear safety spectacles.**

Sodium hydroxide solution is very corrosive. Even when dilute it can damage your eyes. Therefore you **must**:

- **wear safety spectacles.**

Note: Instructions for the preparation of the solutions are summarised on pages 127 to 128. For disposal of residues see page 129.

EXPERIMENT 8.3 Reactions of amines

Requirements per student (or pair)	Notes
6 beakers, 100 cm^3	
2 beakers, 600 cm^3	
4 boiling-tubes, with corks	
stirring rod	
3 teat pipettes	
6 test-tubes, with corks, and test-tube rack	
thermometer, 0–100 °C	
3 watch glasses	
Bunsen burner, tripod, gauze and bench mat	
labels for test-tubes and beakers	
protective gloves and safety spectacles	disposable gloves preferred
spatula	
wood splint	
crushed ice	keep in freezer until asked for
wash-bottle of distilled water	
universal indicator paper	
ammonium chloride, NH_4Cl	~6 g
butylamine, $C_4H_9NH_2$	~5 cm^3
hydrochloric acid, concentrated, HCl	~5 cm^3. Small bottle with dropper placed on a drip tray
naphthalen-2-ol (2-naphthol), $C_{10}H_7OH$	~1 g
phenol, C_6H_5OH	~1 g
phenylamine (aniline) $C_6H_5NH_2$	~5 cm^3
sodium chloride, NaCl	~20 g. Commercial salt will do
sodium nitrite, $NaNO_2$	~2 g
ammonia solution, 2 M NH_3	~5 cm^3. See note below
bromine water, $Br_2(aq)$	~2 cm^3. See note below
copper sulphate solution, 1 M $CuSO_4$	~6 cm^3. Dissolve 25 g of copper(II) sulphate crystals in water and make up to 100 cm^3
lime-water, saturated, $Ca(OH)_2$	~2 cm^3. See note below
sodium hydroxide solution, 2 M NaOH	~20 cm^3. See note below

Note: Instructions for the preparation of the solutions are summarised on pages 127 to 128. For disposal of residues see page 129.

(Please see hazard warnings overleaf.)

EXPERIMENT 8.3 Reactions of amines (continued)

HAZARD WARNING

Phenylamine is **toxic**, by ingestion and by skin absorption.

Phenylamine and butylamine are flammable and give off harmful vapours.

Sodium hydroxide solution is very corrosive. Even when dilute it can damage your eyes. Concentrated hydrochloric acid, phenol and bromine water are corrosive and give off harmful vapours.

Therefore you **must**:

- **wear safety spectacles and gloves;**
- **work in a fume cupboard where possible;**
- **keep bottles away from flames;**
- **keep stoppers on bottles as much as possible.**

EXPERIMENT 8.4 Reactions of aldehydes and ketones

Requirements per student (or pair)	Notes
■ beaker, 250 cm^3	
■ 4 test-tubes	
■ Bunsen burner, tripod, gauze and bench protection mat	
■ protective gloves	disposable type preferred
■ safety spectacles	
■ ethanal (acetaldehyde), CH_3CHO	~10 cm^3
■ propanone (acetone), CH_3COCH_3	~10 cm^3
■ ammonia solution, 2 M NH_3	~5 cm^3. See note below
■ 2,4-dinitrophenylhydrazine solution, $C_6H_3(NO_2)_2NHNH_2$ (Brady's reagent)	~5 cm^3. See note below
■ Fehling's solutions, 1 and 2	~3 cm^3 of each. See note below
■ iodine solution, 10% I_2 in KI(aq)	~1 cm^3. 10 g of iodine and 20 g of potassium iodide in water and make up to 100 cm^3
■ potassium dichromate(VI) solution, 0.1 M $K_2Cr_2O_7$	~2 cm^3. See note below
■ silver nitrate solution, 0.05 M $AgNO_3$	~3 cm^3. See note below
■ sodium hydroxide solution, 2 M NaOH	~10 cm^3. See note below
■ sulphuric acid, dilute, 1 M H_2SO_4	~5 cm^3. See note below

HAZARD WARNING

Ethanal and propanone are irritants to eyes, skin and lungs and are highly flammable. Therefore you **must**:

- **perform the experiment at a fume cupboard;**
- **keep the bottles well away from flames;**
- **keep the stoppers on the bottles as much as possible;**
- **wear safety spectacles and gloves.**

Tollens reagent becomes explosive on evaporation. Therefore you **must**:

- **wash away residues immediately after use.**

Sodium hydroxide solution is very corrosive. Even when dilute it can damage your eyes. Therefore you **must:**

- **wear safety spectacles.**

Fehling's solution 2 is corrosive because it contains sodium hydroxide. It is likely to spurt out of a clean test-tube during heating. Therefore you **must:**

- **wear safety spectacles.**

2,4-dinitrophenylhydrazine is **toxic** (very poisonous). Therefore you **must:**

- **wear safety spectacles and gloves.**

Note: Instructions for the preparation of the solutions are summarised on pages 127 to 128. For disposal of residues see page 129.

EXPERIMENT 8.5 Identifying an unknown carbonyl compound

Requirements per student (or pair)	Notes
Part A	
■ beaker, 250 cm^3	
■ teat pipette	
■ 2 test-tubes	
■ Bunsen burner, tripod, gauze and bench protection mat	
■ protective gloves and safety spectacles	disposable gloves preferred
■ ammonia solution, 2 M NH_3	~2 cm^3. See note below
■ Fehling's solutions, 1 and 2	~2 cm^3 of each. See note below
■ silver nitrate solution, 0.05 M $AgNO_3$	~2 cm^3. See note below
■ sodium hydroxide solution, 2 M NaOH	~5 cm^3. See note below
■ 'unknown' carbonyl compound	~10 cm^3. Consult the teacher
Part B	
■ beaker, 100 cm^3	
■ beaker, 150 cm^3	
■ beaker, 250 cm^3	or steam bath
■ stirring rod	
■ suction filtration apparatus	Büchner or Hirsch funnel, flask, papers, filter pump and tubing
■ filter papers	for drying crystals
■ retort stand, boss and clamp	
■ spatula	
■ crushed ice	on request
■ ethanol, C_2H_5OH	~10 cm^3. IMS is suitable
■ methanol, CH_3OH	~2 cm^3
■ 2,4-dinitrophenylhydrazine solution	~5 cm^3. See note below
■ sulphuric acid, dilute, 1 M H_2SO_4	~2 cm^3. See note below
Part C	
■ *boiling-tube, fitted with cork, stirrer and thermometer (0–360 °C, long-stem)	See diagram overleaf (*or electrical melting-point apparatus)
■ *rubber ring or band	thin section of rubber tubing is suitable
■ *dibutyl phthalate	
■ 3 melting-point tubes	
■ watch glass	

Note: Instructions for the preparation of the solutions are summarised on pages 127 to 128. For disposal of residues see page 129.

(Please see hazard warnings overleaf.)

EXPERIMENT 8.5 Identifying an unknown carbonyl compound (continued)

HAZARD WARNING

Aldehydes, ketones and alcohols are flammable. Therefore you **must**:

- **keep stoppers on bottles of flammable liquids as much as possible;**
- **keep flammable liquids away from flames;**
- **wear safety spectacles.**

Tollens reagent becomes explosive on evaporation. Therefore you **must**:

- **wash away the solution immediately after use.**

Sodium hydroxide solution is very corrosive. Even when dilute it can damage your eyes. Therefore you **must:**

- **wear safety spectacles.**

Fehling's solution 2 is corrosive because it contains sodium hydroxide. It is likely to spurt out of a clean test-tube during heating. Therefore you **must:**

- **wear safety spectacles.**

2,4-dinitrophenylhydrazine is **toxic** (very poisonous). Therefore you **must:**

- **wear safety spectacles and gloves.**

TECHNICIAN'S SHEET

EXPERIMENT 8.6 Reactions of carbohydrates

Requirements per student (or pair)	Notes
General requirements – all parts	
■ beaker, 250 cm^3	for use as a water-bath
■ 6 test-tubes and test-tube rack	
■ Bunsen burner, tripod, gauze and bench mat	
■ safety spectacles	
■ spatula	
■ wash-bottle of distilled water	
■ fructose, $C_6H_{12}O_6$	~1 g
■ glucose, $C_6H_{12}O_6$	~5 g
■ maltose, $C_{12}H_{22}O_{11}$	~1 g. Not essential
■ sucrose, $C_{12}H_{22}O_{11}$	~50 g. Granulated sugar will do
■ starch solution, 2%	~5 cm^3. See below for all recipes
■ sulphuric acid, dilute, 1 M H_2SO_4	~5 cm^3. See note below
Part A	
■ protective gloves	
■ wood splint	
■ sulphuric acid, concentrated, H_2SO_4	~2 cm^3
■ lime-water, saturated, $Ca(OH)_2(aq)$	~5 cm^3. See note below
■ potassium dichromate(VI) solution, 0.1 M $K_2Cr_2O_7$	~5 cm^3. See note below
■ 4 strips of filter paper, 0.5 cm wide	for testing gases
Part B	
■ ammonia solution, 2 M NH_3	~5 cm^3. See note below
■ Fehling's solutions, 1 and 2	~5 cm^3 of each. See note below
■ silver nitrate solution, 0.05 M $AgNO_3$	~5 cm^3. See note below
Part C	
■ 2,4-dinitrophenylhydrazine solution	~10 cm^3. See note below
Part D	
■ litmus paper	
■ ammonia solution, 2 M NH_3	~5 cm^3. See note below
■ Fehling's solutions, 1 and 2	~5 cm^3 of each. See note below
■ iodine solution, 0.1 M I_2 in KI(aq)	~2 cm^3. See note below
Part E	
■ polarimeter, simple type	

Note: Instructions for the preparation of the solutions are summarised on pages 127 to 128. For disposal of residues see page 129.

(Please see hazard warnings overleaf.)

EXPERIMENT 8.6 Reactions of carbohydrates (continued)

HAZARD WARNING

Concentrated sulphuric acid is very corrosive and reacts violently with water, especially when hot. Therefore you **must**:

- **wear safety spectacles and protective gloves;**
- **use small quantities and cool residues before disposal;**
- **pour cold residues slowly into plenty of cold water, stirring to disperse heat.**

Fehling's solution 2 is corrosive because it contains sodium hydroxide. It is likely to spurt out of a test-tube during heating. Therefore you **must**:

- **wear safety spectacles;**
- **ensure that nobody is in line with a test-tube during heating.**

Tollens reagent can become explosive if allowed to evaporate to dryness. Therefore you **must**:

- **wash away residues with plenty of water.**

2,4-dinitrophenylhydrazine is **toxic** (very poisonous). Therefore you **must:**

- **wear safety spectacles and gloves.**

EXPERIMENT 8.7 Observation and deduction exercise 3

Requirements per student (or pair if not used for assessment)	Notes
2 beakers, 100 cm^3	
2 beakers, 250 cm^3	
6 test-tubes and test-tube rack	
Bunsen burner, tripod, gauze and bench mat	
protective gloves and safety spectacles	
spatula	
test-tube holder	
wash-bottle of distilled water	
crushed ice	keep in freezer until required
2 wood splints	
ethanol, C_2H_5OH	~15 cm^3. **Unknown** – labelled C
ethanoic (acetic) acid, glacial, CH_3CO_2H	~15 cm^3. **Unknown** – labelled D
hydrochloric acid, concentrated, HCl	~1 cm^3. Small bottle with dropper placed on a drip tray
phenylammonium chloride (aniline hydrochloride), $C_6H_5NH_3Cl$	~2 g. **Unknown** – labelled E
phenol, C_6H_5OH	~1 g
phosphorus pentachloride, PCl_5	~2 g
potassium iodide, KI	~1 g
sodium carbonate, anhydrous, Na_2CO_3	~5 g
sodium nitrite, $NaNO_2$	~2 g
sulphuric acid, concentrated, H_2SO_4	2 cm^3. Small bottle with dropper placed on a drip tray
ammonia solution, 2 M NH_3	~5 cm^3. See note below
bromine water, $Br_2(aq)$	~2 cm. See note below
lime-water, saturated, $Ca(OH)_2$	~2 cm^3. Issue only on request. See note below
nitric acid, dilute, 2 M HNO_3	~5 cm^3. Carefully add 128 cm^3 of concentrated nitric acid to water and make up to 1000 cm^3
silver nitrate solution, 0.05 M $AgNO_3$	~5 cm^3. See note below
sodium carbonate solution, 1 M Na_2CO_3	~20 cm^3. See Experiment 8.1 – Parts E and F
sodium chlorate(I) solution (hypochlorite), NaClO	~5 cm^3. 10–14% available chlorine as purchased. Commercial bleach is probably adequate
sodium hydroxide solution, 2 M NaOH	~2 cm^3. See note below
potassium dichromate(VI) solution, 0.1 M $K_2Cr_2O_7$	~5 cm^3. See note below

Note: Instructions for the preparation of the solutions are summarised on pages 127 to 128. For disposal of residues see page 129.

(Please see hazard warnings overleaf.)

EXPERIMENT 8.7 Observation and deduction exercise 3 (continued)

HAZARD WARNING

You must wear safety spectacles throughout these tests.

Bromine water, concentrated hydrochloric and sulphuric acids, glacial ethanoic acid, phenol, phosphorus pentachloride, potassium dichromate(VI) solution, silver nitrate solution and sodium hydroxide solution are all corrosive. Therefore you **must**:

- **wear safety spectacles and protective gloves.**

If contact with your skin does occur, wash immediately under a cold tap with plenty of water.

Compounds D and E are not only **toxic** by skin absorption but also, like bromine water, give off harmful vapours. Therefore you **must**:

- **perform any experiments with these compounds at a fume cupboard;**
- **keep the tops on the bottles as much as possible;**
- **wear gloves and safety spectacles.**

Compounds C and D are also flammable so you **must**:

- **keep the bottles away from flames.**

TECHNICIAN'S SHEET

EXPERIMENT 9.1 Determining the molar mass of a gas

Requirements per student (or pair)	Notes
■ volumetric flask, 100 cm^3, with stopper	must be dry
■ thermometer, 0–100 °C	
■ carbon dioxide cylinder with regulator and needle valve	if a CO_2 generator is used (e.g. Kipp's) the gas should be bubbled through water and then concentrated H_2SO_4 to purify it
■ rubber tubing, approximately 90 cm	to connect generator to delivery tube
■ delivery tube, glass, approximately 20 cm	must reach bottom of flask
■ balance, to weigh 100 g to 0.001 g	
■ balance, to weigh 200 g to 0.1 g	must be able to weigh the flask full of water
■ barometer	one reading per class per session, *or* telephone local meteorological office

EXPERIMENT 9.2 Determining the molar mass of a volatile liquid

Requirements per student (or pair)	Notes
gas syringe, 100 cm^3	must be glass and have a free-moving piston
steam generator with safety tube	
steam jacket for gas syringe	see diagram and notes below
thermometer, 0–105 °C	must fit steam jacket
hypodermic syringe, 2 cm^3 and needle	must be glass
silicone rubber, self-sealing	to make temporary seal for needle; e.g. see Griffin catalogue, or bathroom sealant may be suitable
rubber cap, self-sealing	to seal gas syringe nozzle
filter paper	to wipe hypodermic needle
Bunsen burner, tripod, gauze and bench mat	
balance, to weigh to 0.001 g	shared
barometer	one reading per class per session *or* telephone local meteorological office
safety spectacles	
ethyl ethanoate	about 5 cm^3

HAZARD WARNING

- Steam can cause serious scalding.
- Make sure your generator has a safety tube so that pressure cannot build up if the outlet tube becomes blocked. Also, make sure the steam inlet to the jacket is secure and the outlet from the jacket is directed downwards. **Wear safety spectacles.**
- Ethyl ethanoate is flammable. Work well away from the Bunsen burner and do not leave the stopper off the bottle longer than is necessary.

This apparatus may be purchased from laboratory suppliers or made in school. Plastic pipe, such as is used in down-pipes from gutters, can be used with wooden blocks sealing the ends. However, a window must be inserted to allow the volume of gas in the syringe to be measured. Alternatively, a polythene bag can be used with care, provided the arrangement is tested for safety – see *Journal Chem. Ed.* March 1968, volume 45, page 203.

EXPERIMENT 9.3 Determining the molar mass of a gas by effusion

Requirements per student (or pair)	Notes
■ gas syringe, 100 cm^3	must be glass and have a free-moving piston
■ three-way tap	lightly greased to be gas-tight
■ rubber connector, thick walled	to fit tap to syringe
■ aluminium foil, approximately 2 cm^2	cooking foil is ideal
■ quick-setting glue	the teacher may ask you to glue the foil to the tap to seal one exit
■ retort stand, boss and clamp	
■ needle or sharp pin	
■ hydrogen cylinder, regulator and needle valve	if a generator is used instead, the gas must be bubbled through water and then concentrated H_2SO_4 to purify it
■ 2 lengths rubber tubing	to connect three-way tap to hydrogen cylinder and also to domestic gas tap
■ stopclock *or* stopwatch	preferably accurate to 0.1 s

HAZARD WARNING

Hydrogen and domestic gas are not only flammable, but they form explosive mixtures with air. Therefore:

■ **there must be no flames in the laboratory during this experiment.**

EXPERIMENT 9.4 Investigating the hydrolysis of benzenediazonium chloride

Requirements per student (or pair)	Notes
■ beaker, 250 cm^3	
■ glass syringe, 100 cm^3	not greased. Piston must move freely. See diagram overleaf
■ graduated pipette, 5 cm^3	10 cm^3 will serve if necessary
■ measuring cylinder, 10 cm^3	
■ side-arm test-tube, with bung	
■ teat pipette	
■ test-tube	
■ thermometer, 0–100 °C	
■ three-way tap	greased lightly. Must be leakproof
■ crushed ice	~200 cm^3. Enough to cool a test-tube
■ pipette filler	
■ protective gloves	
■ pumice or anti-bumping granules	
■ 2 retort stands, clamps and bosses	
■ rubber tubing, thick wall	2 short pieces to connect three-way tap to syringe and side-arm tube
■ safety spectacles	
■ spatula	
■ stopclock *or* stopwatch	
■ wash-bottle of distilled water	
■ water-bath, thermostatically controlled	if possible, set this up before the lesson and allow to warm up to 40–50 °C
■ hydrochloric acid, concentrated, HCl	~3 cm^3
■ phenylamine, $C_6H_5NH_2$ (aniline)	~1 cm^3
■ sodium nitrite, $NaNO_2$	~1 g

(Please see hazard warnings overleaf.)

EXPERIMENT 9.4 Investigating the hydrolysis of benzenediazonium chloride (continued)

HAZARD WARNING

Phenylamine is **toxic**, by ingestion and by skin absorption. It is also flammable and gives off a harmful vapour.

Concentrated hydrochloric acid is corrosive and gives off a harmful vapour.

Sodium nitrite is an oxidising agent and is **toxic**.

Therefore you **must**:

- **wear safety spectacles and gloves,**
- **work at a fume cupboard where possible,**
- **keep bottles away from flames,**
- **keep stoppers on bottles as much as possible.**

EXPERIMENT 9.5 The kinetics of the reaction between iodine and propanone in aqueous solution

Requirements per student (or pair)	Notes
■ 4 beakers, 100 cm^3	may be shared by several students
■ 4 burettes, 50 cm^3, with stands	(may be shared by several students)
■ 4 funnels, small	(may be shared by several students)
■ 6 test-tubes	preferably optically matched, as supplied for use with colorimeters. Otherwise, select tubes which look identical and fit the colorimeter
■ thermometer, 0–100 °C	
■ 6 bungs *or* corks	to fit the test-tubes
■ colorimeter, with set of filters	
■ safety spectacles	
■ stopclock *or* stopwatch	
■ wash-bottle of distilled water	
■ iodine solution, 0.020 M I_2, in KI(aq)	~30 cm^3. Dissolve 5.08 g of iodine and 16 g of KI in distilled water and make up to 1000 cm^3. Iodine dissolves very slowly – frequent shaking or stirring is necessary to ensure that it dissolves **completely**
■ hydrochloric acid, 2.0 M HCl	~20 cm^3. Add 172 cm^3 of concentrated acid to water and make up to 1000 cm^3
■ propanone solution, 2.0 M CH_3COCH_3 (acetone)	~20 cm^3. Add 147 cm^3 of pure propanone to water and make up to 1000 cm^3

HAZARD WARNING

Concentrated hydrochloric acid is corrosive and gives off a harmful vapour.

Propanone is flammable.

Therefore you **must**:

- **wear safety spectacles and gloves;**
- **keep stoppers on bottles as much as possible;**
- **keep propanone away from flames.**

EXPERIMENT 9.6 Determining the activation energy of a reaction

Requirements per student (or pair)	Notes
■ beaker, 400 cm^3	to be used as a water-bath
■ 4 beakers, 100 cm^3	may be shared by several students
■ 4 burettes, 50 cm^3, with stands	may be shared by several students
■ 4 funnels, small	may be shared by several students
■ 2 boiling-tubes	
■ 2 thermometers, 0–100 °C	
■ Bunsen burner, tripod, gauze and bench mat	
■ retort stand, clamp and boss	
■ safety spectacles	
■ stopclock *or* stopwatch	
■ potassium iodide solution, 0.50 M KI	~25 cm^3. Dissolve 83 g of KI in water and make up to 1000 cm^3
■ potassium peroxodisulphate(VI) (persulphate) solution, 0.020 M $K_2S_2O_8$	~50 cm^3. Dissolve 5.4 g of $K_2S_2O_8$ (or 4.8 g of $Na_2S_2O_8$) in water and make up to 1000 cm^3
■ sodium thiosulphate solution, 0.010 M $Na_2S_2O_3$	~25 cm^3. Dissolve 2.48 g of $Na_2S_2O_3{\cdot}5H_2O$ in water and make up to 1000 cm^3
■ starch solution, 0.2%	~15 cm^3. Mix 0.2 g of 'soluble' starch into a paste with cold water and pour, with constant stirring, into 100 cm^3 of boiling water

EXPERIMENT 9.7 Determining the activation energy of a catalysed reaction

This experiment is a planning exercise for the students, who should work out their own list of requirements and hand it in to you, via the teacher. They will probably request the same apparatus and chemicals as listed for Experiment 9.6, except that the concentration of one (or more) of the solutions may be smaller – preferably the potassium iodide. If in doubt about the suitability of a student's request, consult the teacher. In addition to the requirements for Experiment 9.6, students will need a few cubic centimetres of an iron(III) solution. We suggest the following:

Requirements per student (or pair)	Notes
■ iron(III) chloride solution, 0.05 M $FeCl_3$	~5 cm^3. Dissolve 1.4 g of hydrated solid in water with 5 cm^3 of concentrated hydrochloric acid and make up to 100 cm^3

HAZARD WARNING

Concentrated hydrochloric acid is corrosive and gives off a harmful vapour.

Iron(III) chloride is an irritant.

Therefore you **must**:

- **wear safety spectacles and gloves;**
- **keep stoppers on bottles as much as possible.**

EXPERIMENT 9.8 A bromine 'clock' reaction

Requirements per student (or pair)	Notes
■ 2 beakers, 100 cm^3	
■ 3 beakers, 100 cm^3	may be shared by several students
■ 3 burettes, 50 cm^3, and stands	may be shared by several students
■ 3 funnels, small	may be shared by several students
■ measuring cylinder, 25 cm^3	
■ thermometer, 0–100 °C	
■ safety spectacles	
■ stopclock *or* stopwatch	
■ wash-bottle of distilled water	
■ white tile	
■ methyl orange solution, 0.001%, acidified, labelled 'solution C'	~200 cm^3. Add 20 cm^3 of methyl orange solution (containing 0.5 g dm^{-3}) to 400 cm^3 of 1 M H_2SO_4 and make up to 1000 cm^3
■ methyl orange solution, 0.001%, in 0.40 M KBr, labelled 'solution D'	~100 cm^3. Dissolve 48 g of KBr in water, add 20 cm^3 of methyl orange solution (containing 0.5 g dm^{-3}) and make up to 1000 cm^3
■ phenol solution, 0.0001 M C_6H_5OH	~100 cm^3. Make a 0.010 M solution by dissolving 1.0 g of phenol in 1000 cm^3 of water. Dilute 10 cm^3 of this to 1000 cm^3 immediately before use
■ potassium bromate(V) solution, 0.20 M $KBrO_3$	~100 cm^3. Dissolve 33.4 g of $KBrO_3$ in water and make up to 1000 cm^3
■ potassium bromate(V) solution, 0.0050 M $KBrO_3$	~120 cm^3. Take 25.0 cm^3 of 0.20 M $KBrO_3$ and make up to 1000 cm^3
■ potassium bromide solution, 0.010 M KBr	~120 cm^3. Dissolve 1.2 g of KBr in water and make up to 1000 cm^3
■ sulphuric acid, 0.010 M H_2SO_4	~50 cm^3. Dilute 10 cm^3 of 1.0 M H_2SO_4 to 1000 cm^3. To make 1.0 M H_2SO_4, add 55 cm^3 of concentrated acid, slowly and with constant stirring, to 800 cm^3 of water and make up to 1000 cm^3

HAZARD WARNING

Phenol is corrosive and **toxic** (very dangerous by skin absorption).

Therefore you **must**:

■ **wear safety spectacles and gloves.**

Note: While these solutions need not be made up with great accuracy, it is essential that an individual student has sufficient from the **same stock** to complete each part of the experiment.

BIG MOLECULES

EXPERIMENT 10.1 Chemical properties of carboxylic acids

Requirements per student (or pair)	Notes
■ teat pipette	
■ 5 test-tubes	
■ Bunsen burner and bench mat	
■ filter papers	to blot sodium
■ forceps or tweezers	to handle sodium
■ spatula	
■ protective gloves	
■ safety spectacles	
■ test-tube rack	
■ wash-bottle of distilled water	
■ wood splint	
■ ethanoic (acetic) acid, glacial, CH_3CO_2H	~10 cm^3. **Not** aqueous solution
■ phosphorus pentachloride, PCl_5	~1 g
■ sodium, metal, Na	1 mm cube under the same oil as is used for storage
■ sodium ethanoate (acetate), CH_3CO_2Na	~1 g. The hydrate will serve
■ ammonia solution, 2 M NH_3	~2 cm^3. See note below
■ 2,4-dinitrophenylhydrazine solution, $C_6H_3(NO_2)_2NHNH_2$ (Brady's reagent)	~2 cm^3. See note below
■ iodine solution, 10% I_2 in KI(aq)	~2 cm^3. See note below
■ iron(III) chloride solution, 0.1 M $FeCl_3$	~2 cm^3. Dissolve 2.5 g of solid in water with 2 cm^3 of concentrated hydrochloric acid and make up to 100 cm^3
■ lime-water, saturated, $Ca(OH)_2$	~2 cm^3. See note below
■ sodium hydrocarbonate solution, saturated, $NaHCO_3$	~2 cm^3. Shake about 9 g of solid in 100 cm^3 of water. Decant from excess solid after standing
■ sodium hydroxide solution, 2 M NaOH	~2 cm^3. See note below
■ universal indicator solution	in dropping bottle

HAZARD WARNING

Glacial ethanoic acid is flammable and corrosive.

Sodium hydroxide solution is corrosive.

Sodium and phosphorus pentachloride are extremely reactive, especially with water.

2,4-Dinitrophenylhydrazine (Brady's reagent) is **toxic**.
Therefore you **must**:
- **wear safety spectacles and gloves;**
- **keep stoppers on bottles as much as possible;**
- **keep bottles of flammable liquids away from flames;**
- **keep sodium and phosphorus pentachloride away from water and moist air.**

Note: Instructions for the preparation of the solutions are summarised on pages 127 to 128. For disposal of residues see page 129.

EXPERIMENT 10.2 Chemical properties of ethanoyl chloride

Requirements per student (or pair)	Notes
■ 4 beakers, 100 cm^3	
■ stirring rod	
■ 4 teat pipettes	
■ test-tube	
■ protective gloves	
■ safety spectacles	
■ universal indicator paper	
■ wash-bottle of distilled water	
■ ethanol, C_2H_5OH	~5 cm^3. IMS is suitable
■ ethanoyl (acetyl) chloride, CH_3COCl	~2 cm^3
■ phenylamine (aniline), $C_6H_5NH_2$	~1 cm^3. Fresh sample – see below
■ ammonia solution, 0.880 NH_3	~5 cm^3
■ ammonia solution, 2 M NH_3	~10 cm^3. See note below
■ iron(III) chloride solution, 0.1 M $FeCl_3$	~1 cm^3. See Experiment 10.1
■ sodium carbonate solution, 1 M Na_2CO_3	~1 cm^3. Dissolve 106 g of anhydrous solid in water and make up to 1000 cm^3

HAZARD WARNING

Ethanoyl chloride gives off a vapour which burns the skin, eyes and respiratory tract. Some of its reactions are violent.

Phenylamine is very poisonous if inhaled, swallowed or absorbed through the skin.

Concentrated aqueous ammonia burns the skin and the vapour irritates the eyes.

Ethanol is flammable. Therefore you **must**:
- **wear safety spectacles and gloves;**
- **keep stoppers on the bottles as much as possible;**
- **perform the experiment in a fume cupboard protected by a safety glass.**

Note: Instructions for the preparation of the solutions are summarised on pages 127 to 128. For disposal of residues see page 129.

In this experiment the students add ethanoyl chloride to water, ethanol, ammonia and phenylamine. Since some of these reactions are violent, especially those with ammonia and phenylamine, they **must** perform this in a fume cupboard with the safety glass pulled down as far as is practically possible. In view of the possible hazards the teacher may decide to carry this out as a demonstration. It should be pointed out to students that the concentrated ammonia, 0.880 NH_3 should not be used for part A.3 – only for part C.

Phenylamine Phenylamine (aniline) is colourless when pure, but darkens on storage due to aerial oxidation. It is best to purchase small quantities regularly, but stock that has darkened beyond a clear yellow–brown colour can be purified by distillation in a fume cupboard. Add a little powdered zinc and distil using an air condenser, collecting the fraction at 183–185 °C.

EXPERIMENT 10.3 Preparation of aspirin

Requirements per student (or pair)	Notes
Part A	
■ beaker, 100 cm^3	
■ beaker, 250 cm^3	to act as steam and ice bath
■ conical flask, 100 cm^3	
■ glass rod	
■ 2 measuring cylinders, 10 cm^3 and 100 cm^3	
■ weighing bottle	
■ ground-glass-joint apparatus for reflux	Liebig condenser and 50 cm^3 pear-shaped flask
■ suction filtration apparatus	Büchner or Hirsch funnel, flask, filter papers, filter pump and tubing as in diagram overleaf
■ protective gloves and safety spectacles	
■ retort stand, boss and clamp	
■ spatula	
■ wash-bottle of distilled water	
■ crushed ice	about 125 cm^3
■ access to balance	sensitivity ±0.01 g
■ ethanoic anhydride $(CH_3CO)_2O$	~4 cm^3.
■ 2-hydroxybenzoic acid (salicylic acid)	~11 g
■ phosphoric(V) acid (orthophosphoric acid), 85%	~1 cm^3. Small bottle with dropper placed on a drip tray
Part B	
■ beaker, 100 cm^3	
■ beaker, 250 cm^3	*or* water-bath set at 60 °C
■ specimen bottle with label	
■ glass rod	
■ thermometer, 0–250 °C	
■ Bunsen burner, tripod, gauze and bench mat	
■ filter papers	for drying crystals
■ crushed ice	about 125 cm^3
Part C	
■ *boiling tube fitted with cork, stirrer and long-stem thermometer, 0–250 °C	see diagram overleaf (*or electrical melting-point apparatus)
■ *rubber ring or band	to fix melting-point tubes to the thermometer
■ *retort stand, boss and clamp	
■ *dibutylbenzene-1,2-dicarboxylate (dibutyl phthalate)	~15 cm^3
■ watch glass	
■ 3 melting point tubes	

Note: Instructions for the preparation of the solutions are summarised on pages 127 to 128. For disposal of residues see page 129.

(Please see hazard warnings overleaf.)

EXPERIMENT 10.3 Preparation of aspirin (continued)

HAZARD WARNING

Ethanoic anhydride is corrosive and flammable. It is particularly dangerous to eyes.

2-hydroxybenzoic acid is harmful to eyes, lungs and skin.

Phosphoric(V) acid (orthophosphoric acid) is corrosive and must be kept in a tray to avoid spillage.
Therefore you **must:**

- **wear safety glasses and protective gloves.**

Any aspirin that is prepared must **not** be used medicinally.

EXPERIMENT 10.4 Identification of two organic compounds

Requirements per student (or pair)	Notes
■ beaker, 250 cm^3	
■ stirring rod	
■ 2 teat pipettes	
■ 6 test-tubes	
■ Bunsen burner, tripod, gauze and bench mat	
■ crucible lid	
■ safety spectacles	
■ protective gloves	
■ spatula	
■ test-tube holder	preferably metal
■ test-tube rack	
■ wash-bottle of distilled water	
■ 2 wood splints	
■ 4-oxopentanoic (levulinic) acid solution, 10% $CH_3COCH_2CH_2CO_2H$*	~10 cm^3. Labelled 'F'. Dissolve 10 g of solid in water and make up to 100 cm^3
■ benzaldehyde, C_6H_5CHO	~5 cm^3. Labelled 'G'
■ sodium hydrogencarbonate, $NaHCO_3$	~1 g
■ ammonia solution, 2 M NH_3	~2 cm^3. See note below
■ 2,4-dinitrophenylhydrazine solution, $C_6H_3(NO_2)_2NHNH_2$ (Brady's reagent)	~2 cm^3. See note below
■ lime-water, saturated, $Ca(OH)_2$	on request. See note below
■ potassium iodide solution, 0.6 M KI	~3 cm^3. Dissolve 10 g of solid in water and make up to 100 cm^3
■ silver nitrate solution, 0.05 M $AgNO_3$	~10 cm^3. See note below
■ sodium chlorate(I) (hypochlorite) solution	~10 cm^3. Fresh sample of 10–14% solution. Consult teacher
■ sodium hydroxide solution, 2 M NaOH	~2 cm^3. See note below

HAZARD WARNING

Concentrated hydrochloric acid and solutions of silver nitrate, sodium chlorate(I) (sodium hypochlorite) and sodium hydroxide are all corrosive.

2,4-dinitrophenylhydrazine is **toxic** by skin absorption. Therefore you **must**:

- **wear safety spectacles and gloves throughout.**

Compound G produces a harmful vapour. Therefore you **must**:

- **perform experiments with these compounds at a fume cupboard;**
- **keep tops on bottles as much as possible.**

Compound G is flammable. Therefore you **must**:

- **keep this compound away from flames.**

Note: Instructions for the preparation of the solutions are summarised on pages 127 to 128. For disposal of residues see page 129.

*This may not be available from your usual supplier. Try Sigma-Aldrich Chemical Company Ltd, The Old Brick Yard, New Road, Gillingham, Dorset SP8 4XT. Tel. FREEFONE (0800) 373731 – levulinic acid, L 0626. Price (1997) 50 g £5.30, 250 g £10.90 (£6.50 carriage charge).

EXPERIMENT 10.5 The glycine/copper(II) complex

Requirements per student (or pair)	Notes
■ crystallising dish	
■ filter flask or filter tube	to fit Büchner or Hirsch funnel (see diagram below)
■ stirring rod	
■ test-tube	
■ Büchner funnel, small	a Hirsch funnel may be better, if available
■ filter papers	to fit funnel
■ filter pump and pressure tubing	to fit filter tube or flask
■ safety spectacles	
■ spatula	
■ wash-bottle of distilled water	
■ copper(II) carbonate, $CuCO_3{\cdot}Cu(OH)_2$	~3 g. Finely powdered
■ glycine (aminoethanoic acid), $CH_2NH_2CO_2H$	~1 g

EXPERIMENT 10.6 Biuret test for proteins

Requirements per student (or pair)	Notes
■ teat pipette	
■ 4 test-tubes	
■ safety spectacles	
■ spatula	
■ test-tube rack	
■ wash-bottle of distilled water	
■ 3 protein samples	~0.5 g of solid in 1 cm^3 of solution, e.g. egg albumen, gelatin, fresh milk
■ copper(II) sulphate solution, 0.2 M $CuSO_4$	~2 cm^3. Dissolve 2.5 g of $CuSO_4 \cdot 5H_2O$ in water and make up to 100 cm^3
■ sodium hydroxide solution, 2 M NaOH	~5 cm^3. See note below

HAZARD WARNING

Sodium hydroxide is corrosive. Therefore you **must**:

- **wear safety spectacles.**

Note: Instructions for the preparation of the solutions are summarised on pages 127 to 128. For disposal of residues see page 129.

EXPERIMENT 10.7 Paper chromatography

Requirements per student (or pair)	Notes
■ beaker, 400 cm^3, tall form	chromatography paper must fit inside
■ measuring cylinder, 10 cm^3	
■ 4 melting-point tubes	
■ watch glass, large or cling film	to cover the beaker
■ chromatography paper square, Whatman no. 1, 12.5 cm × 12.5 cm	this paper must not be touched with fingers except at the edges. Wear clean gloves to cut standard 25 cm squares into four and store in a box
■ fume cupboard access	
■ hair dryer	shared by all students
■ oven	to be set at 105–110 °C. Shared by all students
■ 2 paper clips	
■ pencil and ruler	HB or B pencil. 30 cm ruler
■ protective gloves	
■ 2 retort stands, bosses and clamps	
■ safety spectacles	
■ wash-bottle of distilled water	
■ ethanol, C_2H_5OH	~25 cm^3. IMS is suitable
■ ammonia solution, 0.880 NH_3	~5 cm^3
■ aspartic acid solution, 0.01 M	~1 cm^3. Dissolve 0.13 g of solid in 10 cm^3 of propan-2-ol and make up to 100 cm^3 with water
■ leucine solution, 0.01 M	~1 cm^3. Dissolve 0.13 g of solid in 10 cm^3 of propan-2-ol and make up to 100 cm^3 with water
■ lysine solution, 0.01 M	~1 cm^3. Dissolve 0.18 g of lysine hydrochloride in 10 cm^3 of propan-2-ol and make up to 100 cm^3 with water
■ mixture of the three amino acids above	~1 cm^3. Dissolve the masses stated above in 10 cm^3 of propan-2-ol and make up to 100 cm^3 with water
■ ninhydrin solution, aerosol spray	

HAZARD WARNING

Concentrated ammonia solution is corrosive and gives off harmful vapour. Therefore you **must**:

- **wear safety spectacles and gloves;**
- **keep bottles closed as much as possible.**

Ninhydrin sprays give off **toxic** fumes. Therefore you **must**:

- **work at a fume cupboard;**
- **read the instructions on the aerosol can.**

Note: Instructions for the preparation of the solutions are summarised on pages 127 to 128. For disposal of residues see page 129.

EXPERIMENT 10.8 Identification of three organic compounds

Requirements per student (or pair)	Notes
■ access to melting point apparatus	with long-stem thermometer, 0–200 °C
■ beaker, 250 cm^3	to act as water-bath
■ 2 boiling-tubes	with long-stem thermometer, 0–100 °C
■ combustion spoon	
■ spatula	
■ test-tubes	at least 6
■ test-tube rack	
■ measuring cylinder, 10 cm^3	
■ protective gloves	
■ safety spectacles	
■ access to: **unknown** X (benzoic acid) **unknown** Y (propan-2-ol) **unknown** Z (butanone)	only label with letter
■ 2,4-dinitrophenylhydrazine solution, $C_6H_3(NO_2)_2NHNH_2$ (Brady's reagent)	~2 cm^3. See note below
■ potassium dichromate(VI) solution, 0.1 M $K_2Cr_2O_7$	~2 cm^3. See note below
■ sodium carbonate solution, 2 M Na_2CO_3	~20 cm^3. Dissolve 106 g of anhydrous Na_2CO_3 in water and make up to 500 cm^3
■ sulphuric acid, dilute, 2 M H_2SO_4	~5 cm^3. Use 108 cm^3 of concentrated acid and make up to 1 litre with care – see note below

HAZARD WARNING

Compound Y and Z are highly flammable and the vapours are harmful. Therefore you must:

- **keep the compounds away from flames;**
- **perform the experiment in a fume cupboard;**
- **wear safety spectacles and gloves.**

2,4-dinitrophenylhydrazine is **toxic** and 2 M sulphuric acid is corrosive. Therefore you must:

- **wear safety spectacles and gloves;**
- **do boiling point determinations in the fume cupboard.**

Note: Instructions for the preparation of the solutions are summarised on pages 127 to 128. For disposal of residues see page 129.

EXPERIMENT 10.9 Preparing phenyl benzoate

Requirements per student (or pair)	Notes
Part A	
■ conical flask, 250 cm^3, with tight-fitting rubber bung	
■ 2 measuring cylinders, 10 cm^3 and 100 cm^3	
■ weighing bottle	
■ suction filtration apparatus	Büchner or Hirsch funnel, flask, papers, filter pump and tubing. See Experiment 10.3
■ protective gloves and safety spectacles	
■ retort stand, boss and clamp	
■ spatula	
■ wash-bottle of distilled water	
■ access to balance	sensitivity ±0.01 g
■ benzoyl chloride, C_6H_5COCl	~10 cm^3
■ phenol, C_6H_5OH	~5 g
■ sodium hydroxide solution, 2 M NaOH	~100 cm^3. See note below
Part B	
■ boiling-tube	
■ beaker, 250 cm^3	or water-bath
■ specimen bottle, with label	
■ stirring rod	
■ thermometer, 0–100 °C	
■ Bunsen burner, tripod, gauze and bench mat	
■ filter papers	for drying crystals
■ water-bath (thermostatic control)	if available, set at 60 °C
■ crushed ice	
■ ethanol, C_2H_5OH	~25 cm^3. IMS is suitable
Part C	
■ *boiling tube fitted with cork, stirrer and long-stem thermometer, 0–360 °C	see Experiment 10.3 (*or electrical melting-point apparatus)
■ *retort stand, boss and clamp	
■ *rubber ring or band	for fixing melting-point tubes to thermometer
■ *dibutylbenzene-1,2-dicarboxylate (dibutyl phthalate)	~15 cm^3
■ 3 melting point tubes	
■ watch glass	

Note: Instructions for the preparation of the solutions are summarised on pages 127 to 128. For disposal of residues see page 129.

(Please see hazard warnings overleaf.)

EXPERIMENT 10.9 Preparing phenyl benzoate (continued)

HAZARD WARNING

Benzoyl chloride is lachrymatory (produces tears), is irritating to the skin and can cause burns.

Phenol vapour is harmful to the eyes, lungs, and skin. Solid and solution are corrosive and poisonous by skin absorption.

Sodium hydroxide solution is very corrosive. Even when dilute it can damage your eyes.

Ethanol is flammable.
Therefore you **must**:

- **work at a fume cupboard;**
- **wear safety spectacles and protective gloves;**
- **keep the stoppers on bottles;**
- **keep ethanol away from naked flames.**

GROUP IV ELEMENTS

EXPERIMENT 11.1 The reactions of tin and lead and their aqueous ions

Requirements per student (or pair)	Notes
■ 15 test-tubes, and 2 test-tube racks	
■ beaker, 250 cm^3	
■ 3 teat pipettes	
■ Bunsen burner and bench mat	
■ sticky labels for test-tubes	
■ protective gloves and safety spectacles	
■ test-tube holder	
■ wood splints	
■ universal indicator paper	used to test gases
■ lead, Pb	3 or 4 small pieces of foil
■ tin, Sn	2 or 3 small pieces of granulated tin
■ ammonia solution, 2 M NH_3	~20 cm^3. Add 112 cm^3 of 0.880 ammonia to water, and make up to 1000 cm^3
■ hydrochloric acid, dilute, 2 M HCl	~20 cm^3. Add 172 cm^3 of concentrated acid to water and make up to 1000 cm^3
■ hydrochloric acid, concentrated, HCl	~10 cm^3
■ lead(II) nitrate solution, 0.1 M $Pb(NO_3)_2$	~20 cm^3. Dissolve 8.2 g of solid in distilled water and make up to 250 cm^3
■ nitric acid, concentrated, HNO_3	~10 cm^3
■ potassium chromate(VI) solution, 0.1 M K_2CrO_4	~10 cm^3. Dissolve 1.7 g of solid in distilled water and make up to 100 cm^3
■ potassium iodide solution, 0.1 M KI	~5 cm^3. Dissolve 1.7 g of solid in distilled water and make up to 100 cm^3
■ potassium manganate(VII) solution in dilute ethanoic acid, 0.01 M $KMnO_4$ (permanganate)	~10 cm^3. Dissolve 0.16 g of solid in distilled water and make up to 100 cm^3 using about 10 cm^3 of 2 M ethanoic acid
■ sodium hydroxide solution, 2 M NaOH	~20 cm^3. Carefully dissolve 80 g of sodium hydroxide pellets in water and make up to 1000 cm^3
■ sodium sulphide solution, 0.02 M Na_2S	~1 cm^3. Dissolve 0.5 g of solid in distilled water and make up to 100 cm^3. Use the fume cupboard – H_2S may be liberated
■ tin(II) chloride solution, 0.1 M $SnCl_2$	~20 cm^3. See note below

Tin(II) chloride solution, 0.1 M $SnCl_2$ Carefully warm 5.6 g of solid $SnCl_2 \cdot 2H_2O$ with 50 cm^3 of concentrated hydrochloric acid until the solid dissolves. Add two small pieces of tin during the process to discourage oxidation to tin(IV). Make up to 250 cm^3 with distilled water.

(Please see hazard warnings overleaf.)

EXPERIMENT 11.1 The reactions of tin and lead and their aqueous ions (continued)

HAZARD WARNING

Lead compounds are harmful if ingested or absorbed through the skin.

Sodium sulphide is **toxic** and corrosive and evolves highly poisonous hydrogen sulphide gas on contact with acids.

Concentrated hydrochloric acid is very corrosive.

Sodium hydroxide is corrosive, even in 2 M solution.

Concentrated nitric acid is very corrosive and a powerful oxidant.
Therefore you **must**:

- **wear protective gloves and safety spectacles;**
- **use sodium sulphide in the fume cupboard.**

EXPERIMENT 11.2 The preparations and reactions of tin(IV) oxide and lead(IV) oxide

Requirements per student (or pair)	Notes
■ conical flask, 250 cm^3	
■ glass rod	
■ measuring cylinders, 10 cm^3 and 100 cm^3	
■ suction filtration apparatus	use a small Büchner funnel, with correct size filter paper, fitted to a Büchner flask. Attach to water pump
■ 2 specimen tubes	
■ 3 teat pipettes	
■ 10 test-tubes	
■ Bunsen burner, mat, tripod and gauze	
■ evaporating basin	
■ access to balance	
■ safety spectacles and protective gloves	
■ sticky labels	
■ spatula	
■ test-tube holder	
■ wash-bottle of distilled water	
■ wood splints	
■ blue litmus paper	
■ lead (small pieces), Pb	~6 g
■ lead(IV) oxide, PbO_2	~1 g
■ tin (granulated), Sn	~2 g
■ tin(IV) oxide, SnO_2	~1 g
■ hydrochloric acid, dilute, 2 M HCl	~5 cm^3. See Experiment 11.1
■ hydrochloric acid, concentrated, HCl	~5 cm^3
■ nitric acid, dilute, 2 M HNO_3	~30 cm^3. Dilute 120 cm^3 of conc. nitric acid to 1000 cm^3
■ nitric acid, concentrated, HNO_3	~25 cm^3
■ potassium iodide solution in dilute sulphuric acid, 1 M KI	~10 cm^3. Dissolve 16.5 g of the solid in water and make up to 100 cm^3 with distilled water and 10 cm^3 of 2 M H_2SO_4
■ propanone, CH_3COCH_3 (acetone)	~10 cm^3
■ sodium chlorate(I) solution, 1.5 M NaClO (hypochlorite)	~50 cm^3. Use the commercially available solution containing 10–14% chlorate(I)
■ sodium hydroxide solution, 2 M NaOH	~50 cm^3. See Experiment 11.1
■ sodium hydroxide solution, 8 M NaOH	~5 cm^3. Carefully dissolve 32 g of sodium hydroxide pellets in 80 cm^3 of distilled water and make up to 100 cm^3

HAZARD WARNING

Concentrated hydrochloric acid, nitric acid and sodium hydroxide are very corrosive. The vapours of the concentrated acids are also harmful to eyes, lungs and skin.

■ **Take care when making up solutions.**

Concentrated nitric acid is an oxidising agent. Therefore you **must**:

■ **wear safety spectacles;**

■ **avoid contact with skin;**

■ **use these substances in the fume cupboard.**

Nitrogen dioxide and chlorine are **toxic**. Therefore you **must**:

■ **use a fume cupboard when these gases are generated.**

Propanone is very flammable. Therefore you **must**:

■ **keep the stopper on the bottle as much as possible and keep the liquid away from flames.**

EXPERIMENT 11.3 Observation and deduction exercise 4

Requirements per student (or pair)	Notes
■ boiling-tube	
■ filter funnel and filter paper	
■ 10 test-tubes	
■ Bunsen burner and bench mat	
■ litmus paper, blue and red	
■ safety spectacles	
■ spatula	
■ test-tube holder	
■ test-tube rack	
■ wash-bottle of distilled water	
■ wood splints	
■ dilead(II) lead(IV) oxide, Pb_3O_4 (red lead)	~2 g. **Unknown** – labelled Q
■ lead(IV) oxide, PbO_2	~1 g. **Unknown** – labelled I
■ iron(III) chloride solution (neutralised), 0.3 M $FeCl_3$	~5 cm^3. Dissolve 8.1 g of solid in distilled water and make up to 100 cm^3. The solution is neutralised with ammonia solution to the first permanent faint precipitate
■ nitric acid, dilute, 2 M HNO_3	~20 cm^3. See Experiment 11.2
■ potassium chromate(VI) solution, 0.5 M K_2CrO_4	~5 cm^3. See Experiment 11.1
■ potassium iodide solution, 0.1 M KI	~5 cm^3. See Experiment 11.1
■ potassium thiocyanate solution, 0.5 M KSCN	~2 cm^3. Dissolve 5.0 g of solid in distilled water and make up to 100 cm^3
■ silver nitrate solution, 0.02 M $AgNO_3$	~2 cm^3. Dissolve 0.30 g of solid in distilled water and make up to 100 cm^3
■ sodium hydroxide solution, 2 M NaOH	~5 cm^3. See Experiment 11.1
■ sulphuric acid, dilute, 2 M H_2SO_4	~5 cm^3. Carefully add 110 cm^3 of concentrated sulphuric acid, slowly and with constant stirring to 800 cm^3 of distilled water and make up to 1000 cm^3
■ tin(II) chloride solution, 0.25 M $SnCl_2$	~30 cm^3. **Unknown** – labelled H. See Experiment 11.1 but use 14 g of solid $SnCl_2{\cdot}2H_2O$ with 50 cm^3 of concentrated hydrochloric acid and make up to 250 cm^3 with distilled water

HAZARD WARNING

Sodium hydroxide solution is corrosive, even at 2 M concentration. Therefore you **must**:

- **wear safety spectacles throughout, especially when making up solutions.**

Powder Q is harmful if ingested.

EXPERIMENT 11.4 Illustrating the oxidation states of vanadium

Requirements per student (or pair)	Notes
■ conical flask, 100 cm^3	
■ filter funnel	
■ measuring cylinder, 25 cm^3	
■ 5 test-tubes	
■ Bunsen burner, tripod, gauze and bench mat	
■ filter papers	
■ safety spectacles	
■ protective gloves	
■ spatula	
■ test-tube holder	
■ test-tube rack	
■ wash-bottle of distilled water	
■ ammonium polytrioxovanadate (ammonium metavanadate), NH_4VO_3	~1 g
■ sodium sulphite, Na_2SO_3	~1 g
■ sulphuric acid, concentrated, H_2SO_4	~10 cm^3
■ zinc, powder, Zn	~2 g. Small granules may be used, but powder is better. Avoid dust – it is dangerously flammable
■ potassium iodide solution, 0.05 M KI	~2 cm^3. Dissolve 0.8 g of solid in water and make up to 100 cm^3
■ potassium manganate(VII) (permanganate) solution, 0.02 M $KMnO_4$	~10 cm^3. Dissolve 3.2 g of solid in water and make up to 1000 cm^3
■ sodium thiosulphate solution, 0.1 M $Na_2S_2O_3$	~2 cm^3. Dissolve 25 g of the pentahydrate in water and make up to 1000 cm^3
■ sulphuric acid, dilute, 1 M H_2SO_4	~25 cm^3. Carefully add 55 cm^3 of concentrated acid, slowly and with constant stirring, to 800 cm^3 of distilled water and make up to 1000 cm^3

HAZARD WARNING

Concentrated sulphuric acid is corrosive and reacts violently with water. Therefore you **must**:

- **wear safety spectacles and gloves;**
- **when diluting, add acid to water, not water to acid;**
- **mop up small spillages with excess water.**

EXPERIMENT 11.5 Illustrating the oxidation states of manganese

Requirements per student (or pair)	Notes
■ filter funnel	
■ stirring rod	
■ 6 test-tubes	
■ 3 filter papers	
■ safety spectacles	
■ protective gloves	
■ spatula	
■ test-tube rack	
■ wash-bottle of distilled water	
■ manganese(IV) oxide, MnO_2 (manganese dioxide)	~1 g
■ manganese(II) sulphate-4-water, $MnSO_4{\cdot}4H_2O$	~1 g. The monohydrate will do
■ sulphuric acid, concentrated, H_2SO_4	~1 cm^3
■ sodium hydroxide solution, 2 M NaOH	~10 cm^3. Carefully dissolve 80 g of pellets in distilled water and make up to 1000 cm^3
■ potassium manganate(VII) (permanganate) solution, 0.01 M $KMnO_4$	~15 cm^3. Dilute the solution made for Experiment 11.4
■ sulphuric acid, dilute, 1 M H_2SO_4	~5 cm^3. See Experiment 11.4

HAZARD WARNING

Sodium hydroxide is corrosive, even in dilute solution. Therefore you **must**:

- **wear safety spectacles and gloves when making up solutions.**

Concentrated sulphuric acid is corrosive and reacts violently with water. Therefore you **must**:

- **wear safety spectacles and gloves;**
- **when diluting, add acid to water, not water to acid;**
- **mop up small spillages with excess water.**

EXPERIMENT 11.6 Relative stabilities of some complex ions

Requirements per student (or pair)	Notes
■ 6 test-tubes	
■ test-tube rack	
■ safety spectacles	
■ protective gloves	
■ wash-bottle of distilled water	
■ copper(II) sulphate solution, 0.20 M $CuSO_4$	~5 cm^3. Dissolve 50 g of the pentahydrate in water and make up to 1000 cm^3
■ 1,2-diaminoethane solution, 0.10 M $(CH_2NH_2)_2$	~25 cm^3. Dissolve 6.0 g of solid in water and make up to 1000 cm^3
■ edta solution (disodium salt), 0.10 M $C_{10}H_{14}O_8N_2Na_2$	~25 cm^3. Dissolve 37 g of the dihydrate in water and make up to 1000 cm^3
■ sodium chloride solution, 5 M NaCl	~25 cm^3. Shake 300 g of solid with water and make up to 1000 cm^3. This solution is almost saturated
■ sodium ethanedioate (oxalate) solution, 0.20 M $(CO_2Na)_2$	~25 cm^3. Dissolve 2.7 g of solid in water and make up to 1000 cm^3

HAZARD WARNING

Sodium ethanedioate is **toxic**. Therefore you **must**:

■ **wear safety spectacles and disposable gloves.**

EXPERIMENT 11.7 Determining the formula of a complex ion

Requirements per student (or pair)	Notes
■ 3 beakers, 100 cm^3	
■ 3 burettes, 50 cm^3	these may be shared with other students. 10 cm^3 graduated pipettes, with safety fillers, may be substituted
■ 3 funnels, small	for filling burettes (if used)
■ 8 test-tubes, with bungs *or* corks	optically matched, if possible, and fitting the colorimeter. If these tubes hold less than 15 cm^3, students will need 8 larger tubes **as well**
■ 3 burette stands	if burettes are used
■ colorimeter, with filters	
■ safety spectacles	
■ test-tube rack	
■ wash-bottle of distilled water	
■ ammonia solution, 0.10 M NH_3	~60 cm^3. See note below
■ ammonium sulphate solution, 2 M $(NH_4)_2SO_4$	50 cm^3. Dissolve 264 g of solid in water and make up to 1000 cm^3
■ copper(II) sulphate solution, 0.10 M $CuSO_4$	~25 cm^3. See note below

HAZARD WARNING

Copper(II) sulphate is harmful (poisonous) and the dust irritates lungs and eyes.

0.880 ammonia solution is corrosive and gives off an irritant vapour.
Therefore you **must:**

- **wear eye protection and gloves when making up solutions.**

Ammonia solution and copper(II) sulphate solution It is important for this experiment that these two solutions should be of **equal concentration**, but not necessarily 0.10 mol dm^{-3} precisely. We suggest that you do not rely on old stock solutions, but standardise the ammonia against a known acid. The copper sulphate solution can then be made to the **same** concentration.

If you need to make fresh ammonia solution, add 5.6 cm^3 of 0.880 ammonia to water, make up to 1000 cm^3 and mix thoroughly before standardisation.

0.10 M $CuSO_4$ requires 25.0 g of the pentahydrate per 1000 cm^3. Adjust this concentration as necessary.

EXPERIMENT 11.8 Some redox chemistry of copper

Requirements per student (or pair)	Notes
■ beaker, 250 cm^3	
■ boiling-tube	
■ 4 test-tubes	
■ Bunsen burner and bench mat	
■ safety spectacles	
■ spatula	
■ test-tube holder	must fit boiling-tube
■ test-tube rack	
■ wash-bottle of distilled water	
■ copper turnings, Cu	~1 g
■ copper(II) chloride-6-water, $CuCl_2{\cdot}6H_2O$	~3 g. If this is not available, a mixture of sodium chloride and copper(II) sulphate may be used
■ sodium chloride, NaCl	~3 g
■ sodium sulphite, Na_2SO_3	~1 g. The hydrated form may be used
■ sodium thiosulphate, $Na_2S_2O_3$	~1 g. The hydrated form may be used
■ copper(II) sulphate solution, 0.1 M $CuSO_4$	~2 cm^3. See Experiment 11.7
■ potassium iodide solution, 0.1 M K1	~4 cm^3. Dissolve 1.7 g of solid in water and make up to 100 cm^3
■ Fehling's solutions, 1 and 2	~3 cm^3

HAZARD WARNING

Fehling's solution 2 contains sodium hydroxide and is therefore corrosive.

■ **Wear safety spectacles and handle with care, especially when heating.**

TECHNICIAN'S SHEET

EXPERIMENT 11.9 Investigating the use of cobalt(II) ions as catalyst

Requirements per student (or pair)	Notes
■ beaker, 250 cm^3	
■ measuring cylinder, 25 cm^3	
■ stirring rod	
■ teat pipette	
■ 2 test-tubes	
■ Bunsen burner, tripod, gauze and bench mat	
■ safety spectacles	
■ protective gloves	
■ spatula	
■ test-tube rack	
■ thermometer, 0–100 °C	
■ wash-bottle of distilled water	
■ cobalt(II) chloride, $CoCl_2{\cdot}6H_2O$	~1 g
■ hydrogen peroxide solution, 1.7 M H_2O_2	~20 cm^3 of '20-volume' H_2O_2, as purchased
■ potassium sodium 2,3-dihydroxybutanedioate (tartrate), $CO_2K(CHOH)_2CO_2Na{\cdot}4H_2O$	~1 g. Commonly known as 'Rochelle salt'

HAZARD WARNING

Hydrogen peroxide is corrosive and an oxidant.
Handle with care.

EXPERIMENT 11.10 Catalysing the reaction between iodide ions and peroxodisulphate ions

Requirements per student (or pair)	Notes
■ 4 beakers, 100 cm^3	
■ boiling-tube	
■ 4 burettes, 50 cm^3	may be shared with other students
■ conical flask, 150 cm^3	
■ 4 funnels, small	for filling burettes
■ 4 test-tubes	
■ safety spectacles	
■ stopclock	
■ test-tube rack	
■ wash-bottle of distilled water	
■ chromium(III) chloride solution, 0.1 M $CrCl_3$	~1 cm^3. Dissolve 3 g of the hexahydrate in water and make up to 100 cm^3
■ iron(II) sulphate solution, 0.1 M $FeSO_4$	~1 cm^3. Dissolve 3 g of the heptahydrate in water and make up to 100 cm^3
■ iron(III) chloride solution, 0.1 M $FeCl_3$	~1 cm^3. Dissolve 3 g of the hexahydrate in water with 2 cm^3 of conc. HCl and make up to 100 cm^3
■ manganese(II) sulphate solution, 0.1 M $MnSO_4$	~1 cm^3. Dissolve 2.5 g of the tetrahydrate in water and make up to 100 cm^3
■ potassium chromate(VI) solution, 0.1 M K_2CrO_4	~1 cm^3. Dissolve 2 g of solid in water and make up to 100 cm^3
■ potassium manganate(VII) (permanganate) solution, 0.1 M $KMnO_4$	~1 cm^3. Dissolve 1.5 g of solid in water and make up to 100 cm^3
■ potassium iodide solution, 0.2 M KI	~40 cm^3. Dissolve 33 g of solid in water and make up to 1000 cm^3
■ potassium peroxodisulphate (persulphate) solution, 0.2 M $K_2S_2O_8$	~30 cm^3. Dissolve 54 g of solid in water and make up to 1000 cm^3
■ sodium thiosulphate, 0.01 M $Na_2S_2O_3$	~40 cm^3. See Experiment 11.4
■ sulphuric acid, dilute, 1 M H_2SO_4	~25 cm^3. See Experiment 11.4
■ starch solution, 0.2%	~20 cm^3. Mix 0.2 g of 'soluble' starch into a paste with cold water and pour, with constant stirring, into 100 cm^3 of boiling distilled water

The six 0.1 M solutions should be supplied in dropping bottles if possible. Otherwise, six teat pipettes will be needed.

EXPERIMENT 11.11 Observation and deduction exercise 5

Requirements per student (or pair)	Notes
■ beaker, 100 cm^3	
■ 3 boiling-tubes	
■ 4 test-tubes	
■ 2 teat pipettes	
■ watch glass	
■ Bunsen burner and bench mat	
■ flame-test rod or wire	
■ litmus papers, red and blue	or pH papers
■ safety spectacles	
■ spatula	
■ test-tube rack	
■ wash-bottle of distilled water	
■ wood splints	
■ ammonium polytrioxovanadate (metavanadate), NH_4VO_3	~0.5 g in a separate tube, labelled 'substance C'
■ potassium dichromate(VI), $K_2Cr_2O_7$	~2 g in a separate tube, labelled 'substance D'
■ copper, powder, Cu	~1 g
■ hydrochloric acid, concentrated, HCl	~5 cm^3
■ hydrogen peroxide solution, 1.7 M H_2O_2	~5 cm^3 of '20 volume' H_2O_2
■ zinc, powder, Zn	~1 g. Avoid zinc **dust**, which is dangerously flammable
■ potassium iodide solution, 0.6 M KI	~5 cm^3. Dissolve 10 g of solid in water and make up to 100 cm^3
■ sodium hydroxide solution, 2 M NaOH	~50 cm^3. See Experiment 11.5
■ sodium thiosulphate solution, 0.1 M $Na_2S_2O_3$	~10 cm^3. Dissolve 25 g of the pentahydrate in water and make up to 1000 cm^3
■ sulphuric acid, dilute, 1 M H_2SO_4	~20 cm^3. See Experiment 11.4

HAZARD WARNING

Concentrated hydrochloric acid, sulphuric acid and sodium hydroxide are all corrosive. Concentrated hydrochloric acid gives off an irritant vapour and concentrated sulphuric acid reacts violently with water.
Therefore you **must:**

■ **wear eye protection and gloves when making up solutions.**

Hydrogen peroxide is corrosive and is a powerful oxidant.
Therefore you **must:**

■ **wear eye protection and gloves, especially if you dilute '100 volume' solution.**

Potassium dichromate(VI) is an oxidant and an irritant.

■ **Handle with care, avoiding inhalation of dust.**

P-BLOCK ELEMENTS

EXPERIMENT 12.1 Reactions of aluminium

Requirements per student (or pair)	Notes
■ beaker, 100 cm^3	
■ filter funnel	
■ 10 test-tubes	
■ Bunsen burner and bench mat	
■ filter papers	
■ safety spectacles	
■ protective gloves	
■ test-tube holder	
■ test-tube rack	
■ wash-bottle of distilled water	
■ aluminium foil, Al	6 pieces, 3 cm × 3 cm. Kitchen foil is suitable
■ litmus papers, red and blue	
■ copper(II) chloride solution, 0.1 M $CuCl_2$	~10 cm^3. Dissolve 1.7 g of $CuCl_2{\cdot}2H_2O$ in water and make up to 100 cm^3
■ hydrochloric acid, dilute, 2 M HCl	~10 cm^3. Pour 172 cm^3 of concentrated hydrochloric acid slowly into 750 cm^3 of distilled water and make up to 1000 cm^3
■ mercury(II) chloride solution, 0.1 M $HgCl_2$	a few drops. Dissolve 2.7 g of solid in water and make up to 100 cm^3
■ sodium carbonate solution, 1 M Na_2CO_3	~20 cm^3. Dissolve 106 g of anhydrous Na_2CO_3 in water and make up to 1000 cm^3
■ sodium hydroxide solution, 2 M NaOH	~20 cm^3. Carefully dissolve 80 g of sodium hydroxide pellets in water and make up to 1000 cm^3
■ sulphuric acid, dilute, 1 M H_2SO_4	~20 cm^3. Carefully add 55 cm^3 of concentrated sulphuric acid, slowly and with constant stirring, to 800 cm^3 of water and make up to 1000 cm^3

HAZARD WARNING

Mercury(II) chloride is **toxic** by ingestion and skin absorption. Copper(II) chloride is harmful (poisonous) and dust irritates eyes and lungs.

■ **Avoid ingestion and wear safety spectacles and gloves.**

Concentrated hydrochloric acid is corrosive and gives off an irritating vapour.

■ **Make up dilute acid with care, wearing gloves and eye protection.**

Mercury(II) chloride is toxic by ingestion and skin absorption.

■ **Avoid dust and wear gloves and eye protection.**

Sodium hydroxide and concentrated sulphuric acid are corrosive and generate heat when dissolved in water.

■ **Wear eye protection and make up solutions with great care.**

EXPERIMENT 12.2 Reactions of the oxoacids of nitrogen and their salts

Requirements per student (or pair)	Notes
■ beaker, 250 cm^3	
■ 2 boiling-tubes	
■ 6 test-tubes, with corks *or* bungs to fit	
■ Bunsen burner and bench mat	
■ protective gloves	
■ safety spectacles	
■ spatula	
■ wash-bottle of distilled water	
■ wood splints	
■ aluminium foil, Al	3 cm square of kitchen foil
■ copper turnings, Cu	4–5
■ Devarda's alloy (Cu 50%, Al 45%, Zn 5%)	~1 g
■ ice, crushed	keep in freezer until requested
■ iron(II) sulphate, $FeSO_4{\cdot}7H_2O$	~1 g
■ magnesium ribbon, Mg	~5 cm
■ sodium nitrite, $NaNO_2$	~3 g
■ nitric acid, concentrated, 16 M HNO_3	~4 cm^3
■ nitric acid, dilute, 2 M HNO_3	~10 cm^3. Carefully add 128 cm^3 of concentrated nitric acid to 800 cm^3 of water and make up to 1000 cm^3
■ potassium iodide solution, 0.5 M KI	~1 cm^3. Dissolve 8.3 g of KI crystals in water and make up to 100 cm^3
■ potassium manganate(VII) (permanganate) solution, 0.02 M $KMnO_4$	~1 cm^3. Dissolve 3.2 g of $KMnO_4$ crystals in water and make up to 100 cm^3
■ sodium hydroxide solution, 2 M NaOH	~15 cm^3. See Experiment 12.1
■ sulphuric acid, dilute, 1 M H_2SO_4	~20 cm^3. See Experiment 12.1

HAZARD WARNING

Sodium nitrite is **toxic** and an oxidant (assists fire).

■ **Handle with care.**

Potassium manganate(VII) is a powerful oxidant.

■ **Handle with care and keep well away from concentrated sulphuric acid, with which it reacts explosively.**

Concentrated nitric acid, sodium hydroxide and concentrated sulphuric acid are corrosive and generate heat when dissolved in water.

■ **Wear eye protection and make up solutions with great care.**

EXPERIMENT 12.3 Investigating reactions of the oxosalts of sulphur

Requirements per student (or pair)	Notes
■ bottle labelled 'silver residues'	
■ filter funnel	
■ 10 test-tubes	
■ boiling-tube	
■ Bunsen burner and bench mat	
■ filter papers and test strips	strips are for testing gases
■ safety spectacles	
■ protective gloves	
■ spatula	
■ test-tube holder	
■ test-tube rack	
■ wash-bottle of distilled water	
■ wood splints	
■ sodium peroxodisulphate (persulphate), $Na_2S_2O_8$	~1 g. The potassium salt will serve
■ sodium sulphate-10-water, $Na_2SO_4{\cdot}10H_2O$	~1 g
■ sodium sulphite-7-water, $Na_2SO_3{\cdot}7H_2O$	~1 g
■ sodium thiosulphate-5-water, $Na_2S_2O_3{\cdot}5H_2O$	~1 g
■ sulphur, finely powdered roll, S	~1 g. Preferred to 'flowers'
■ hydrochloric acid, dilute, 2 M HCl	~15 cm^3. See Experiment 12.1
■ iodine solution, 0.2 M I_2 in KI(aq)	10 cm^3. Dissolve 5.1 g of iodine and 10 g of KI crystals in water and make up to 100 cm^3
■ iron(III) chloride solution, 0.5 M $FeCl_3$	5 cm^3. Dissolve 13 g of solid in water with 2 cm^3 of concentrated hydrochloric acid and make up to 100 cm^3
■ potassium dichromate(VI) solution, 0.1 M $K_2Cr_2O_7$	~2 cm^3. Dissolve 2.9 g of solid in water and make up to 100 cm^3
■ potassium iodide solution, 0.5 M KI	~5 cm^3. See Experiment 12.2
■ silver nitrate solution, 0.1 M $AgNO_3$	~10 cm^3. Dissolve 1.7 g of solid in water and make up to 100 cm^3. Keep in amber bottles
■ sodium hydroxide solution, 2 M NaOH	~10 cm^3. See Experiment 12.1
■ sodium peroxodisulphate (persulphate) solution, 0.2 M $Na_2S_2O_8$	~5 cm^3. 4.8 g $Na_2S_2O_8$ or 5.4 g $K_2S_2O_8$ per 100 cm^3
■ sodium sulphate solution, 0.2 M Na_2SO_4	~5 cm^3. 6.5 g $Na_2SO_4{\cdot}10H_2O$ per 100 cm^3
■ sodium sulphite solution, 0.2 M Na_2SO_3	~15 cm^3. 5.0 g $Na_2SO_3{\cdot}7H_2O$ per 100 cm^3
■ sodium thiosulphate solution, 0.2 M $Na_2S_2O_3$	~5 cm^3. 5.0 g $Na_2S_2O_3{\cdot}5H_2O$ per 100 cm^3

(Please see hazard warnings overleaf.)

EXPERIMENT 12.3 Investigating reactions of the oxosalts of sulphur (continued)

HAZARD WARNING

Sodium peroxodisulphate is an oxidant.

■ **Handle with care.**

Sodium sulphite is harmful (poisonous).

■ **Wear eye protection.**

Concentrated hydrochloric acid is corrosive and gives off an irritating vapour.

■ **Wear eye protection and make up solutions with care.**

Iodine is corrosive.

■ **Wear gloves and eye protection when making up solutions.**

Potassium dichromate(VI) is an oxidant.

■ **Handle with care.**

Silver nitrate is corrosive and **toxic**.

■ **Wear eye protection when making up solutions.**

Sodium hydroxide is corrosive and generates heat when dissolved in water.

■ **Wear eye protection and make up solutions with care.**

EXPERIMENT 12.4 Anodising aluminium

Requirements per student (or pair)	Notes
■ 2 beakers, 100 cm^3	
■ Bunsen burner, tripod, gauze and bench mat	
■ cotton-wool	
■ forceps or tweezers	
■ 2 pins, large	heavy gauge, as used in optics experiments or entomology
■ protective gloves	
■ safety spectacles	
■ thermometer, 0–100 °C	
■ 2 pieces of aluminium sheet	7 cm × 3 cm. Mounted on wooden bars as shown below
■ ammeter, 1 A	
■ d.c. supply, 12 V	
■ 4 connecting leads	2 must have crocodile clips at one end
■ rheostat, 10 Ω, 4 A	
■ sulphuric acid, dilute, 1 M H_2SO_4	~75 cm^3. See Experiment 12.1
■ propanone, CH_3COCH_3	~2 cm^3
■ dye solution	a few drops or, if convenient, ~50 cm^3. Dissolve 0.2 g of alizarin red or methyl violet in 100 cm^3 of water and add two drops of 2 M NaOH. Congo red, and some other dyes, may be used but require 20% ethanol for solution. Standard indicator solutions are not sufficiently concentrated

HAZARD WARNING

Concentrated sulphuric acid is corrosive and generates much heat when dissolved in water. Make up solutions with great care. You **must**:

- **wear gloves and safety spectacles.**

Propanone is highly flammable and has a harmful vapour.

- **Keep the bottle away from flames and replace stoppers as soon as possible.**

EXPERIMENT 12.5 Observation and deduction exercise 6

Requirements per student (or pair)	Notes
■ beaker, 100 cm^3	
■ boiling-tube	
■ evaporating dish	
■ flame-test wire or rod	platinum wire or silica rod
■ 4 test-tubes in rack	
■ watch glass	
■ Bunsen burner, tripod, gauze and bench mat	
■ safety spectacles	
■ protective gloves	
■ filter paper strips	
■ litmus paper, red and blue	
■ wash-bottle of distilled water	
■ iron(II) sulphate, $FeSO_4 \cdot 7H_2O$	~2 g
■ ammonium ethanoate (acetate) solution, 3 M $CH_3CO_2NH_4$	~10 cm^3. Dissolve 23 g of solid in water and make up to 100 cm^3
■ barium chloride solution, 0.5 M $BaCl_2$	~2 cm^3. Dissolve 12 g of crystals in water and make up to 100 cm^3
■ copper(II) sulphate solution, 0.5 M $CuSO_4$	~5 cm^3. Dissolve 12 g of hydrate in water and make up to 100 cm^3
■ hydrochloric acid, concentrated, 12 M HCl	~5 cm^3
■ hydrochloric acid, dilute, 2 M HCl	~15 cm^3. See Experiment 12.1
■ lead(II) ethanoate (acetate) solution, 0.25 M $(CH_3CO_2)_2Pb$	~2 cm^3. Dissolve 9.5 g of hydrate in water and make up to 100 cm^3
■ potassium dichromate(VI) solution, 0.5 M $K_2Cr_2O_7$	~2 cm^3. Dissolve 14 g of solid in water and make up to 100 cm^3
■ potassium iodide solution, 0.5 M KI	~10 cm^3. See Experiment 12.2
■ potassium thiocyanate solution, 0.5 M KSCN	~2 cm^3. Dissolve 5 g of solid in water and make up to 100 cm^3
■ silver nitrate solution, 0.1 M $AgNO_3$	~4 cm^3. See Experiment 12.3
■ sodium chloride solution, 0.1 M NaCl	~4 cm^3. Dissolve 0.6 g of solid in water and make up to 100 cm^3
■ sodium peroxodisulphate (persulphate) solution, 0.05 M $Na_2S_2O_8$, labelled 'solution D'	~10 cm^3. Dissolve 1.2 g of $Na_2S_2O_8$ (or 1.4 g of $K_2S_2O_8$) in water and make up to 100 cm^3
■ sodium sulphate solution, 0.1 M Na_2SO_4 labelled 'solution B'	~15 cm^3. Dissolve 3.4 g of $Na_2SO_4 \cdot 10H_2O$ in water and make up to 100 cm^3
■ sodium thiosulphate solution, 0.1 M $Na_2S_2O_3$, labelled 'solution C'	~40 cm^3. Dissolve 22 g of $Na_2S_2O_3 \cdot 5H_2O$ in water and make up to 1000 cm^3
■ sulphuric acid, dilute, 1 M H_2SO_4	see Experiment 12.1

HAZARD WARNING

Barium chloride, copper(II) sulphate, lead ethanoate and potassium thiocyanate are harmful (poisonous).

■ **Avoid dust and make up solution with care.**

Concentrated hydrochloric acid is corrosive and gives off an irritating vapour.

■ **Wear eye protection and gloves.**

Potassium dichromate(VI) and potassium peroxodisulphate are oxidants.

■ **Handle with care.**

Silver nitrate is corrosive and **toxic**.

When making up solution you **must:**

■ **wear eye protection.**

Results Table 4.11

Test	Observations	Inferences
a Heat approximately 0.1 g of F in a Pyrex tube, at first gently and then more strongly, until the change is complete. Cool and keep the residue. Test any gases evolved		
b Make an aqueous solution of the residue from **a** and carry out the following tests on portions: i) Add aqueous silver nitrate followed by dilute nitric acid ii) Add aqueous chlorine iii) Add aqueous lead(II) ethanoate (lead acetate)		

Results Table 4.12

Substance provided	Name of non-metal	Name and formula of substance/ion	Oxidation number
F			

Results Table 4.13 Tests with unknown substance D

Method	Observations	Inferences
a		
b		
c		

Results Table 4.14 Tests with unknown substance E

Method	Observations	Inferences
a		
b		
c		
d		
e		

 © JOHN MURRAY

Results Table 4.15 Experiments to test inferences

Inference tested	Test and observations	Conclusion

Results Table 8.7

Test	Observations	Inferences
1. **a** Place 1 cm^3 of C in a test-tube and add an equal volume of water. Now add a little anhydrous sodium carbonate **b** Repeat test 1**a** using D		
2. (The reactions 2**a** and 2**b** should be performed at a fume cupboard) **a** Place 2 or 3 cm^3 of C in a dry beaker. Add a little phosphorus pentachloride (**care**) **b** Repeat test 2**a** using D (**care**)		
3. Mix, in a small beaker, about 2 cm^3 of each of C, D and concentrated sulphuric acid (**care**). Warm gently but do not boil. Pour the mixture into an excess of aqueous sodium carbonate in a large beaker. Smell the product		
4. Mix about 5 cm^3 of D with an equal volume of aqueous potassium dichromate in a test-tube. Pour the mixture into about 10 cm^3 of nearly boiling dilute sulphuric acid in a small beaker. Smell the mixture		
5. In a small beaker dissolve a few crystals of potassium iodide in about 10 cm^3 of D. Add a few drops of aqueous sodium hydroxide. Now add 2 or 3 cm^3 of aqueous sodium chlorate(I) (hypochlorite). Warm the mixture but do not boil. Cautiously smell the products		
6. Make a solution of E in distilled water and use portions for the following tests: **a** To 2–3 cm^3 of the solution add aqueous bromine **b** To 2–3 cm^3 of the solution add aqueous silver nitrate. Then add dilute nitric acid. Finally add dilute aqueous ammonia		
7. Dissolve a little E in about 1 cm^3 of concentrated hydrochloric acid and dilute to about 4 cm^3 with distilled water. Cool the tube in an ice–water mixture and add a few drops of aqueous sodium nitrite (to be prepared by dissolving sodium nitrite in distilled water). Leave the tube in the ice–water mixture. Dissolve a few crystals of phenol (**care**) in 7–8 cm^3 of aqueous sodium hydroxide, cool this solution, and then add it to the cold solution prepared as above		

 © JOHN MURRAY

Results Table 8.8

	Reasoning
Functional group in C	
Functional group in D	

Results Table 10.4

Experiment	Observations	Inferences
a To 2 cm^3 of the solution of F add 2,4-dinitrophenylhydrazine reagent		
b Prepare a sample of Tollens reagent as follows: to 5 cm^3 of aqueous silver nitrate in a test-tube add one to two drops of aqueous sodium hydroxide. Then add dilute aqueous ammonia until only a trace of precipitate remains. Now add five drops of the solution of F and place the tube in hot water. (Pour the contents of the tube down the sink on completion of the test)		
c Add a little of the solution of F to some sodium hydrogencarbonate		
d To 1 cm^3 of the solution of F add 3 cm^3 of aqueous potassium iodide and then 10 cm^3 of sodium chlorate(I) (sodium hypochlorite) solution		
e Place **one** drop of G on an inverted crucible lid and ignite G from above		
f To 1–2 cm^3 of G add 2,4-dinitrophenylhydrazine reagent		
g Prepare a sample of Tollens reagent as in **b**. Add five drops of G, shake the mixture and place the tube in hot water. (Pour the contents of the tube down the sink on completion of the test)		

 © JOHN MURRAY

Results Table 10.6

Test	Observations	Inferences
1. **a** Burn a small amount of X in a combustion spoon **b** Shake a small amount of X in a test-tube with 5 cm^3 of water. First try cold water then hot. (Use a hot water-bath) **c** To a small amount of X in a test-tube add 5 cm^3 of 0.5 mol dm^{-3} sodium carbonate solution **d** Determine the melting point of X		
2. **a** Burn a small volume of Y in a combustion spoon **b** Shake a small amount (1–2 cm^3) of Y with 5 cm^3 of water in a test-tube. First try cold water then hot. (Use a hot water-bath) **c** To a few drops of Y in a test-tube, add about 5 cm^3 of a mixture of aqueous potassium dichromate(VI) and 2M H_2SO_4 and warm in a hot water-bath **d** Determine the boiling point of Y in a fume cupboard		
3. **a** Burn a small volume of Z on a combustion spoon **b** Shake a small amount (1–2 cm^3) of Z with 5 cm^3 of water in a test-tube. First try cold then hot water **c** Add a few drops of Z to 5 cm^3 of a solution of 2,4-dinitrophenyl-hydrazine (Brady's reagent) in a test-tube **d** Determine the boiling point of Z in a fume cupboard		

Results Table 10.7

Compound	Functional group	Reasoning
X		
Y		
Z		

 © JOHN MURRAY

Data for Experiment 10.8 (Part B) Identification of three unknown compounds

Quantitative analysis for three unknown compounds

X C 68.9%, H 4.9%, O 26.2%
Y C 60.0%, H 13.3%, O 26.7%
Z C 66.7%, H 11.1%, O 22.2%

a The mass spectrum of compound X. **b** The infra-red spectrum of compound X.

a

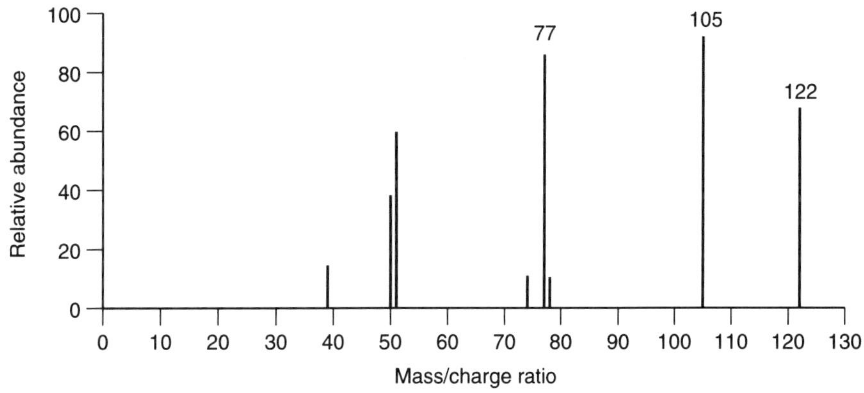

b

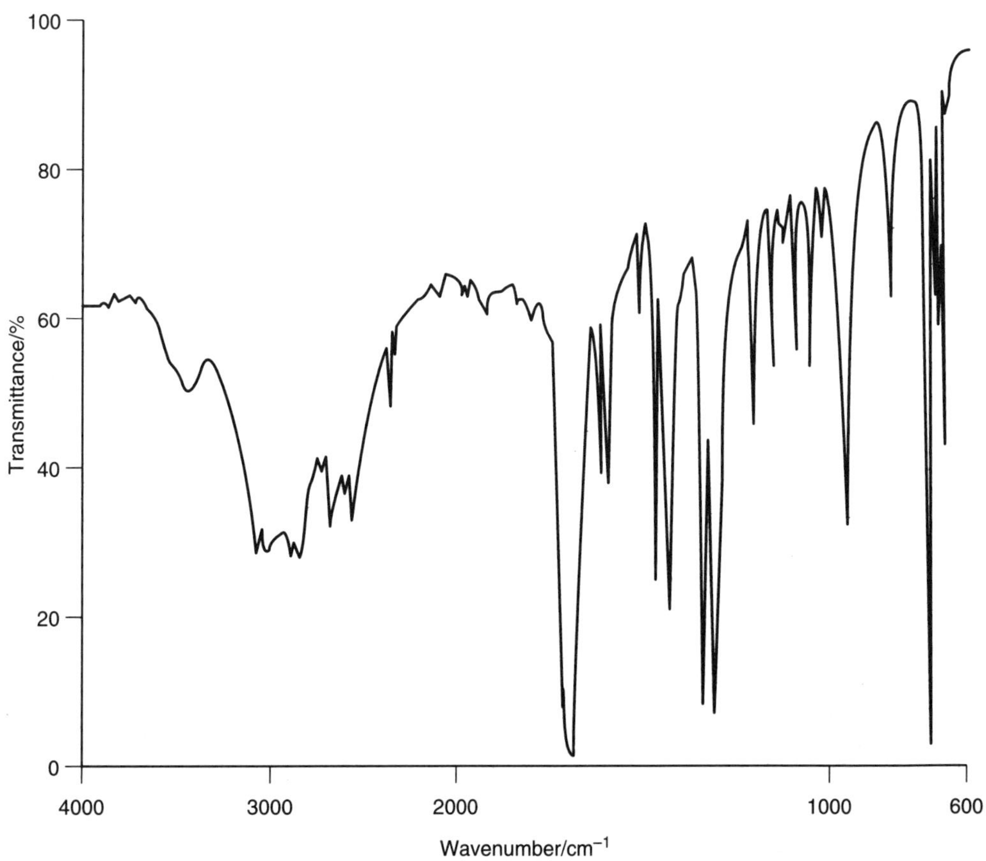

a The mass spectrum of compound Y. **b** The infra-red spectrum of compound Y.

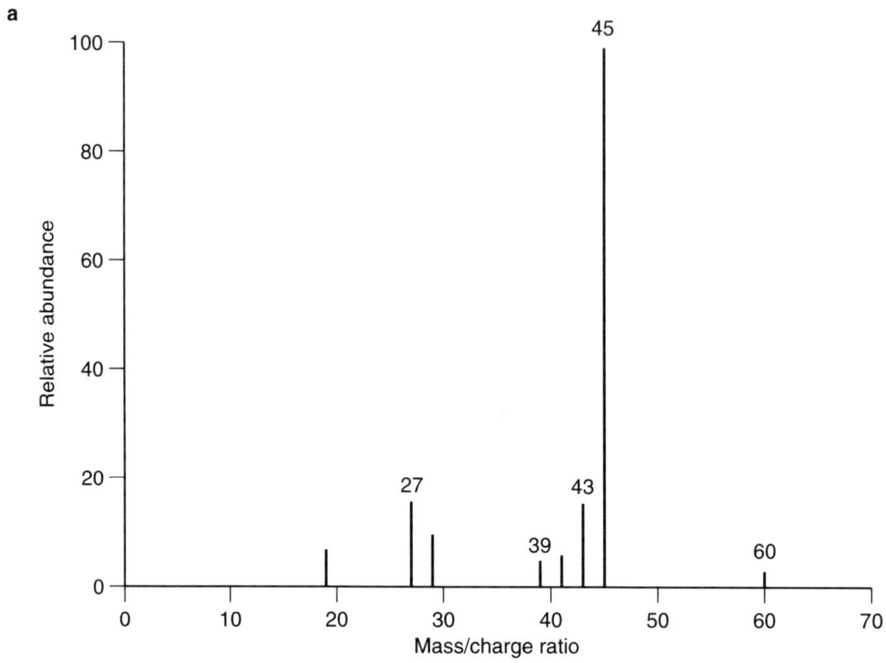

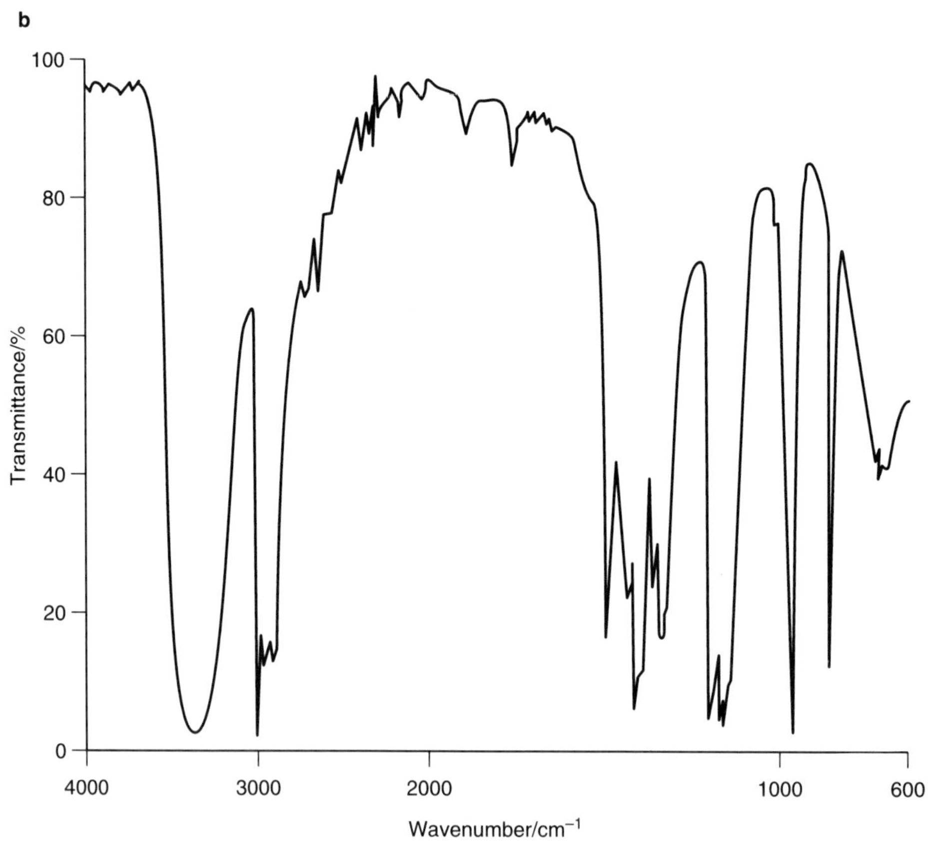

a The mass spectrum of compound Z. **b** The infra-red spectrum of compound Z.

a

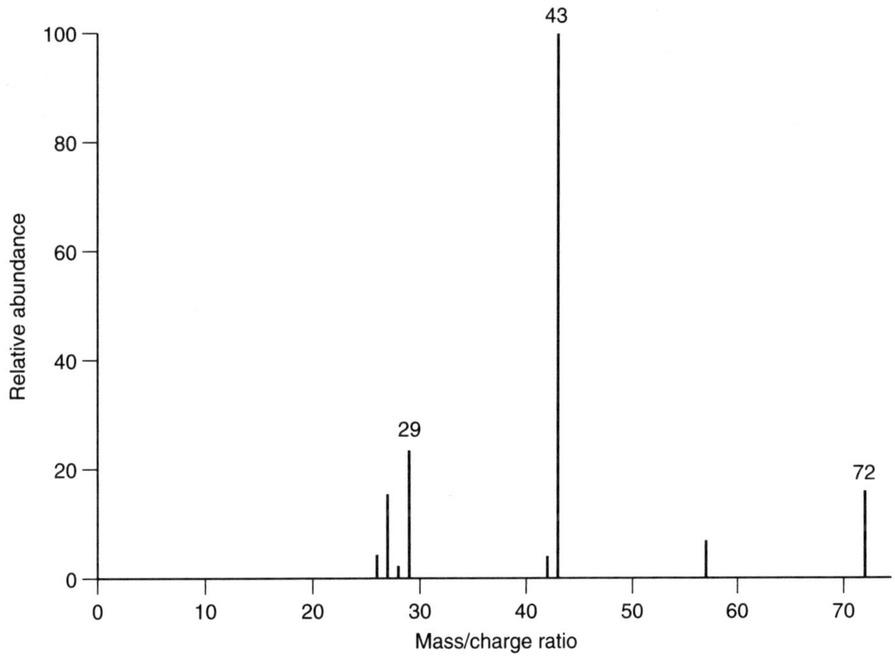

b

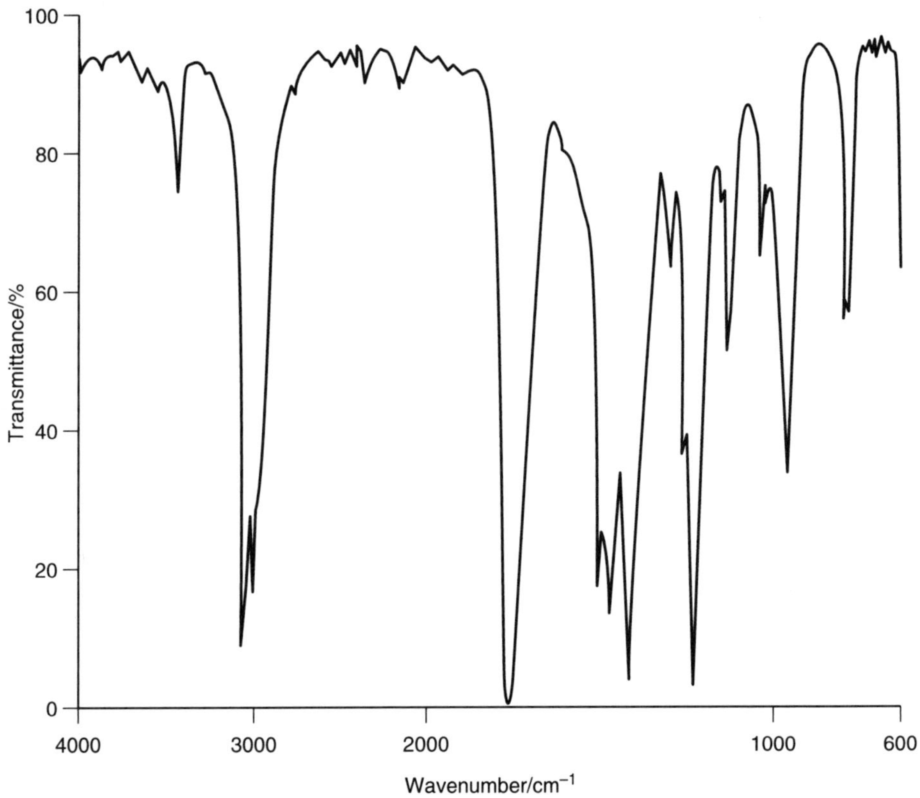

Results Table 11.5
Tests on Q

Method	Observations	Inferences
1.		
2.		
3.		

 © JOHN MURRAY

Results Table 11.6
Experiment to test inference

Inference	Test and observations	Conclusion

Results Table 11.7
Tests on H and I

Test	Observations	Inferences
1. To 1 cm^3 of the solution of H add aqueous silver nitrate followed by dilute nitric acid		
2. To 1 cm^3 of the solution of H add aqueous sodium hydroxide until in excess		
3. To 1 cm^3 of aqueous iron(III) chloride add a few drops of aqueous potassium thiocyanate. To this solution add some of the solution of H		
4. Heat some of solid I in a Pyrex boiling-tube. Allow to cool. Add 8–10 cm^3 of dilute nitric acid to the residue and boil the mixture for one or two minutes. Filter if necessary and use portions of the cool solution for the following tests: **a** To 1 cm^3 of the solution add aqueous sodium hydroxide **b** To 1 cm^3 of the solution add dilute sulphuric acid **c** To 1 cm^3 of the solution add aqueous potassium chromate(VI)		

Results Table 11.13

Test	Observations	Inferences
1. Warm three-quarters of your sample of C with 4–5 cm^3 of aqueous sodium hydroxide		
2. Add approximately 10 cm^3 of dilute sulphuric acid to the remainder of your sample of C in a boiling-tube and warm the mixture in order to dissolve the solid. Use portions of the solution for the following tests:		
a To 1–2 cm^3 of the solution of C add an equal volume of aqueous potassium iodide. Then add, dropwise, aqueous sodium thiosulphate until there is no further change		
b To 2 cm^3 of the solution add a little copper powder and warm		
c To 3–4 cm^3 of the solution add a little zinc powder and allow the mixture to stand. It is suggested that you make observations for about five minutes and then allow the mixture to stand for a further 30 minutes, making observations from time to time. During this time you should proceed with other tests		
3. Carry out a flame-test on substance D		
4. Dissolve some D in the minimum quantity of distilled water and use portions of the solution for the following tests:		
a Add a few drops of this solution to a mixture of 1–2 cm^3 of aqueous potassium iodide with an equal volume of dilute sulphuric acid. Then add aqueous sodium thiosulphate dropwise		
b To 1 cm^3 of the solution of D add aqueous sodium hydroxide. Then add dilute sulphuric acid until in excess		
c To about 2 cm^3 of the solution of D add an equal volume of dilute sulphuric acid. Then add hydrogen peroxide solution drop-wise until there is no further change		(No inference required)
d Transfer the solution from **c** to a boiling-tube and add about 10 cm^3 of aqueous sodium hydroxide and 1 cm^3 of hydrogen peroxide solution. Heat the resulting solution		

Results Table 12.6

Test	Observations	Inferences
1. To 2 or 3 cm^3 of solution B add a few drops of dilute hydrochloric acid. Now add a few drops of aqueous barium chloride		
2. To 2 or 3 cm^3 of solution B add a few drops of aqueous lead ethanoate (lead acetate). Now add an excess of aqueous ammonium ethanoate (ammonium acetate) and shake the mixture		
3. To about 5 cm^3 of solution C in a boiling tube add about 10 cm^3 of dilute hydrochloric acid. Allow the mixture to stand for a minute or two. Cautiously smell the mixture. Warm, if necessary, and test the gas evolved. Describe below how you performed this test **Method**		
4. **a** In a small beaker place 2 or 3 cm^3 of aqueous copper(II) sulphate. Acidify with two drops of dilute sulphuric acid. Now add about 5 cm^3 of aqueous potassium iodide **b** To the mixture obtained in 4 **a** add solution C until in excess, swirling well		
5. **a** In a small beaker mix about 2 cm^3 of aqueous sodium chloride with an equal volume of aqueous silver nitrate **b** To the mixture obtained in 5 **a** add solution C until in excess, swirling well		
6. Dissolve about 1 g of iron(II) sulphate crystals in dilute sulphuric acid and add a few drops of aqueous potassium thiocyanate. Now add a few drops of solution D		
7. **a** In a small beaker place about 5 cm^3 of aqueous potassium iodide and a few drops of dilute sulphuric acid. Now add about 5 cm^3 of solution D. Warm this mixture **b** To the mixture obtained in 7 **a** add solution C until in excess, swirling well		
8. Evaporate a small quantity of solution B to dryness and perform a flame test on the residue. Describe below how you do this **Method**		